The evidence is becoming ever clearer that continued use of fossil fuels as the world's dominant energy supply is damaging the environment and causing changes in global climate patterns. Fossil fuels are also a finite resource, and the current situation is, therefore, unsustainable: future use of alternative methods of energy supply is inescapable. The promise of renewable energy can only be realized through significant investments in research and development on alternative, sustainable technologies.

Prospects for Sustainable Energy explores the historical origins, technical features, marketability, and environmental impacts of the complete range of alternative sustainable energy technologies: solar, biomass, wind, hydropower, geothermal power, ocean energy sources, solar-derived hydrogen fuel, and the energy storage technologies necessary to operate them competitively. Arguments for and against implementation of each option are addressed, and the book makes a technological and economic assessment of the market readiness of these technologies and the prospects of each for reaching a competitive status.

The aim of this book is to inform policy analysts and decision makers of the options available for sustainable energy production. The book is written to be accessible to an audience from a broad range of backgrounds and scientific training. It will also be a valuable supplementary text for advanced courses in environmental studies, energy economics and policy, and engineering.

EDWARD S. CASSEDY is Professor Emiritus of Electrical Engineering at the Polytechnic University, Brooklyn, New York. He is the author of numerous research papers in applied physics, and energy economics and policy, and co-author of *Introduction to Energy* (with Peter Grossman, first edition published in 1990, second edition 1998; Cambridge University Press).

For Bernice

Prospects for Sustainable Energy

A Critical Assessment

Edward S. Cassedy

CAMBRIDGE
UNIVERSITY PRESS

CAMBRIDGE UNIVERSITY PRESS
Cambridge, New York, Melbourne, Madrid, Cape Town, Singapore, São Paulo

Cambridge University Press
The Edinburgh Building, Cambridge CB2 2RU, UK

Published in the United States of America by Cambridge University Press, New York

www.cambridge.org
Information on this title: www.cambridge.org/9780521631204

First published 2000
Reprinted 2001
This digitally printed first paperback version 2005

A catalogue record for this publication is available from the British Library

Library of Congress Cataloguing in Publication data

Cassedy, Edward S.
Prospects for sustainable energy : a critical assessment / Edward S. Cassedy.
    p.   cm.
Includes index.
ISBN 0–521–63120–3
1. Renewable energy sources. 2. Energy policy. I. Title.
TJ808.C37 1999
333.79′4 – dc21 99–11969 CIP

ISBN-13 978-0-521-63120-4 hardback
ISBN-10 0-521-63120-3 hardback

ISBN-13 978-0-521-01837-1 paperback
ISBN-10 0-521-01837-4 paperback

Contents

Preface   *page* vii

I Introduction   1

     Introduction   3

II Candidate technology assessment   17

     1   Solar energy sources   19
     2   Biomass energy   67
     3   Windpower   113
     4   Hydroelectric power   136
     5   Energy storage   153
     6   Geothermal energy   188
     7   Ocean energy   193
     8   Nuclear fusion   200
     9   Hydrogen fuel from renewable resources   207

III The prospects for technological change toward sustainability   223

     10 Summary assessment of the technologies   225
     11 Research and development   231

     Appendices   259

Appendix A *Energy cost analysis*   261
Appendix B *Glossary*   264
Appendix C *The conduct and management of research and development*   269

Index   279

Preface

The purpose of this book is to assess the prospects of producing significant amounts of the world's energy needs from renewable resources – alternatives to fossil fuels. The assessments will be technological, economic, and social in nature.

The underlying basis for judgment on my part, in all three areas, is my experience in a lifetime of research, engineering, and teaching, having been a member of the faculty in electrical engineering at the Polytechnic University for 35 years and a staff member in two different R&D laboratories for almost 10 years before that.

In recent years, I have participated in engineering/economic projects to examine energy storage and solar heat for industrial use. I also, during the years of the "energy crisis", taught a course to graduate engineering and management students in energy policy issues. Finally, in the 1990s, I have taught graduate students and guided doctoral research in electric power system economics and planning.

There is, moreover, another experience of my recent years which has helped to shape this book. This was the teaching and writing leading to the publication of *Introduction to Energy – Resources, Technology and Society*, Cambridge University Press, 1st edn, 1990, 2nd edn, 1998; this was co-authored with P. Z. Grossman (now a professor of economics at Butler University). It was written by us as part of a new academic program, one of several across the country called Science, Technology and Society (STS) programs, and was dedicated to fostering *technological literacy* amongst students not majoring in any branch of science or engineering. The objective for these students was to lay a foundation of understanding in science sufficient for them to engage in critical analysis of the problems of our technological world.

I found from this experience – in short – that the two worlds of C. P. Snow did not have to be entirely separate. This was then the starting point for me to presume that I could write a book such as the present one, addressing technical issues for an audience that would be technologically literate but not necessarily technically expert. I, therefore, gained the confidence that the critical aspects of the prospects for alternative energy production could be conveyed to economists, business people, and policy analysts who are technologically literate but do not have the technical expertise of science or engineering training.

My experience in academic science and engineering includes another realm besides that of technological understanding, however, one that is well recognized as influencing the processes of scientific progress and technological change. This is the realm of social behavior, as it appears in the world of academia, research, and technology, molding the outcomes from that world. It has been during the course of the R&D efforts that I first observed many of the phenomena of social behavior that are summarized in Chapter 11 (Research and development) and Appendix C (The conduct and management of research and development). Also, many of the social conditioning aspects of doctoral training mentioned in the latter chapter have been evident at close hand in the course of administering the graduate program in electrical engineering at the University.

In Appendix C, I have used these experiences and observations as a guide to search the literature of the behavioral sciences – mostly, in the sociology and history of science. The behavioral aspects of research and development are important in the assessment of prospects of new technologies in two major areas, in my estimation. In popular terms, I would call these "group think" and "boosterism." Regarding the first, the behavioral literature which I have cited gives ample evidence of the "corporate/consensual" nature of the scientific establishment and how this tends to channel the paths of inquiry in scientific disciplines. Regarding the second, my way of dealing with the seemingly inevitable promotional tendencies in R&D literature has been simply to omit it from my text whenever it seems to crop up in my references. As I explain in the Introduction, my practice will be to cite the present status of each technology and what the prospects for progress are *but* pointedly to avoid repeating the optimistic projections of those who have a stake in those outcomes.

This all may not win favor in some quarters, nor will the critical nature of some of my assessments in some cases. It has long been my feeling, however, that the needs of public officials and private investors are ill served by the technological optimism and, often, out and out boosterism that the technical community feeds them about new technologies. Even though most of these promising messages are not blatant distortions or patently self serving, being more the product of the culture of the technical community, they are still misleading to those without expertise among policy makers, investment advisors, and the public at large. I take a position along with Ken Karas, Past President of the American Wind Energy Association, who cautioned his wind-industry colleagues about issuing overly optimistic projections for their technology. I favor sustainable technologies and advocate their development, but I shrink from overly optimistic advocacy. In fact, that is what this book is all about. My sense is that independent, realistic assessment, rather than boosterism, does a better service for the continued support of R&D programs for sustainable energy technologies.

I am indebted to several people for their comments and guidance in earlier drafts of this book, including Peter Meier (Idea, Inc., Washington, DC), Peg Reese and Victor Rezendes (US General Accounting Office, Washington, DC), and my former colleague Peter Grossman (now at Butler University, Indianapolis, IN). Certain anonymous reviewers also added measurably to the quality of the book. My editor, Matt Lloyd, has been of immeasurable help with his advice and support of this effort. I offer my thanks to Ms Carletta Lino, of Polytechnic University, who typed the earliest version of the book. I am also grateful to the following individuals who gave of their time and care in supplying me with many of the figures and photographs used here: Dona McLain of ASEA, Martin Filian of Hydro Quebec, Jerry Dominquez of So. Calif. Edison, Michelle Montague and Kent Stuart of AWEA, Adam Mickevicius of Morrison Knudsen, Clay Aldrich of Siemans Solar, Chris Hocker of Consolidated Hydro, Rich Hayes of UCS, Sarah Tuchio of the Ford Motor Co., Vickie Kourkouliotio of NREL, and Diane Lear of Voith Hydro. In addition, Jerry Tuskin (NREL) and George Lof (Lof Energy Systems) offered helpful comments along with supplying their figure materials. Finally, I want to thank my wife, Bernice, for her loving support and patience through several years of my closeting myself with this effort.

Introduction

Introduction

The necessity for assessment

What are the prospects for energy supplies in the 21st century for this country and the world at large? Energy, after all, drives the industrialized world and is essential to the development of the Third World. At the same time, energy production is the chief pressure for the degradation of the environment in the industrial world and promises to add to the myriad problems of the developing world as well. We are prompted to ask: "Are there alternatives to the present path of the world's energy, economy and environment?".

Fossil-fueled combustion accounts in the late 1990s for 89% of the commercial energy consumption in the USA (Bodansky, 1991; Weisel & Kelly, 1991) and 80% of the total energy use worldwide (Hollander,1990). The worldwide fossil-fuel emissions of the greenhouse gas carbon dioxide have been estimated at over 20 billion metric tonnes (mt) annually. The International Energy Agency (IEA) (Ferrier, 1996, 1997; IEA, 1996) expects world primary energy demand to increase with an annual growth rate of around 2%, reaching 50% above the levels of the 1990s by the second decade of the 21st century. They also forecast that 90% of that energy will still be supplied by fossil fuels in that decade. It is to be expected that energy-driven carbon dioxide emissions will increase in about the same proportions. The IEA further expects that the largest developing countries (China and India) will have a greater increase in CO_2 emissions than the already industrialized nations of the Organization for Economic Cooperation and Development (OECD). And while control of the emissions of pollutants, such as sulfur dioxide (SO_2) and carbon fine particles, is being established in many countries, there remains a debate over further reductions. Even in the face of optimistic energy-efficiency scenarios, Anderson (1995) expects that the growth of demand from developing countries will result in a net growth of world fossil-fuel consumption into the 21st century.

These facts alone tell the tale of environmental pressures. There is an increasing consensus in scientific circles that anthropogenic sources of the "greenhouse gases" (carbon dioxide, methane, chlorofluorocarbons (CFCs), nitrous oxide, etc.) are contributing to climate change worldwide (NAS, 1991; Rosen & Glasser, 1992; Steen, 1994). (Of these, CO_2 and nitrous oxide are emissions from fuel combustion.) The documented rise in global average temperatures is now undisputed. The steady annual increases in atmospheric concentrations of CO_2 and the other gases have been recorded and have long been unquestioned. The theoretical understanding of the mechanism of interaction with thermal radiation, whereby heat is held by these gases in the upper atmosphere, is unchallenged. Whereas the atmospheric physics issues of counterbalancing effects (e.g., cloud cover) and the dynamics of atmospheric and oceanic circulations have not been fully resolved, there has been increasing scientific agreement that the prospect of global warming resulting from human activities is real, with fossil-fuel combustion a major contributor. But in the face of these findings, there is still not a concerted effort by the industrial nations to take strong measures to reduce greenhouse gas emissions worldwide (*New York Times*, 1997). Nonetheless, in view of these prospects, there has been a call to "decarbonize the world economy" (Goldemberg, 1996).

In addition, however, are the prospects of long-term exhaustibility of the fossil resources and their present uneven geographical distribution. The industrial world has already experienced the implications of the latter in the second half of the 20th century. The "energy crises" of the 1970s were simply disruptions in the supplies of oil for the industrial nations who were, and still are, heavily dependent on oil imports. The question of *security* of supplies for oil-importing nations could be revisited by the turn of the century, according to some analysts (Amirahmadi, 1995; Ferrier, 1996, 1997; Salameh, 1996; Goldstein, 1997). It is expected that the USA, which has a major share of the world market, will have to import over two thirds of its oil in the first decade after the turn of the century, as its domestic production continues to decline. Other oil-importing industrial countries, such as in Western Europe, are apt to fall into greater dependence on imports, with North Sea production declining into the second decade of the 21st century and limited prospects in the future for imports from Eastern Europe. Japan will have to compete with a growing demand by China for oil supplies, which may still be dependent on imports from the Persian Gulf, even with increased exploration and production in the Far East (IEA, 1995).

By the second decade of the 21st century, it is expected that the Organization of Petroleum Exporting Countries (OPEC) could be supplying 50% or more of the world oil market, at a rate of 50 million barrels per day or greater (IEA, 1996). But this will be technically feasible only if they have made the investments necessary for that production capacity. In addition, the political situation in the Persian Gulf shows no signs of stabilization within the foreseeable future (Salameh, 1996). This will make prospects precarious for uninterrupted exports for the major fraction of OPEC production and require extensive diplomatic and military efforts by the industrial powers to prevent further disruptions.

There is even a running debate about the exhaustability of world oil and natural gas resources. Industry spokesmen continue to assert that technology gains in exploration will tend to increase world reserves (*Oil & Gas Journal*, 1996); this has the support of some analysts (Cleveland & Kaufman, 1997) who claim that the historic decline in exploration yield per unit of drilling is beginning to slow. Other analysts, however, challenge the optimistic view that innovations in exploration are delaying the exhaustion of these resources (Laherrere & Perrodon, 1997). They believe that peaks of world production of oil will be not far off the peak predicted in the historic logistics curves of Hubbert, produced as long ago as the beginning of the 20th century, resulting in an increase in prices thereafter. They also predict that world gas production will peak in the second decade of the 21st century, thus flying in the face of current optimism about the future of natural gas, as promoted by the gas interests.

The present commitment to fossil-fuel energy is best exemplified by the massive use of oil: the world consumption is over 25 billion barrels annually and the USA uses close to 6 billion barrels. Two thirds of this demand is for transportation, which is over 95% petroleum based in the US. Similarly for coal, the US consumption is currently approaching 1 billion metric tonnes annually, about 85% of which supplies nearly 55% of the nation's electricity. World coal consumption is expected to increase to about 3 billion tons annually by the year 2010, up from 2.3 billion in 1990 (IEA, 1996). Annual wholesale expenditures for all fossil fuels (including natural gas) in the USA exceed $250 billion ($250B). The aggregate value of capital assets in fossil-fueled technology, including power plants, petroleum-fueled vehicles, industrial plants and residential/commercial space heating, exceeds two trillion dollars in the USA. Similar consumption and investments for fossil-fired technologies occur in other industrialized countries (Ager-Hanssen, 1993).

If we examine the energy prospects for the less-developed countries (LDCs), we find a potential demand several times that of the industrialized world, since the LDCs total population is over three times as great. Already, the demand for energy is growing in the Third World at about 7% per annum, compared with around 3% for the developed world (Hollander, 1990), and much of this demand is being met with imported oil. This continues despite the constraints imposed on the oil-importing LDCs by the debts of the oil-crisis period of the 1970–80s (Cassedy & Grossman, 1990).

With the further thrusts toward development, large demands for fossil fuels can be expected unless alternative energy technologies become available. These demands will be for oil in those LDCs having resources or able to afford imports. Other LDCs can, and have started to, develop other fossil fuels, such as coal and natural gas, wherever these are indigenous resources (Cassedy & Meier, 1988). This will accelerate the trends in global pollution and climate change. However, those LDCs without access to energy resources will be crippled in their attempts to start industrial development. Technological developments in alternative energy sources will help to alleviate these pressures on the LDCs but cannot be assumed, for the most part, to originate within those countries. Technological innovations will most likely emanate from the research and development (R&D) establishments of the industrial countries, and our focus for assessing the prospects for these developments will accordingly be there, in the USA, Europe, and Japan.

In recent years, advocates from the industrialized world have promoted the use of renewable resources (such as biomass or hydropower) or inexhaustible resources (such as solar energy or wind power) as the solution to the development dilemmas of the Third World. Some have even proposed that such resources be the chief means of powering the developing world, in order that it does not repeat the despoliation that has accompanied industrialization of the Northern nations. As we shall see in later chapters, the technical, financial, and institutional barriers to alternative technologies make widespread non-fossil development quite uncertain in the near future, in the developed world as well as in the LDCs. The developing nations have already indicated their impatience with unproven solutions and have escalated their use of fossil fuels as their financing and development programs have permitted.

Overall, it is becoming apparent that the demands for energy, both for the continued prosperity of the industrial nations and for the development of the non-industrial nations, are coming more and more into conflict with the world environment. Such conclusions are not confined to environmental activists; the differences expressed in public debate seem only to be about the degree and urgency of this conflict. Indeed, the questions posed for public policy seem, to a large degree, to be about priorities: the economy or the environment? And this, like most questions of public policy, has come down to a matter of values and perceived worth.

Some industrial countries have made major commitments to slowing the growth of the greenhouse gases, as part of the world agreements for *sustainable development* (Brundtland, 1987). Several countries in the European Community (EC) have put emphasis on developing alternative source technologies, such as the "renewables" (solar, wind, and biomass) (Chartier *et al.*, 1996). Denmark, Holland, and Germany have declared policies for development of these new technologies and have active research programs working toward their goals. However, the policies of most governments in the industrialized world, including the USA, have emphasized merely the promotion of energy efficiency in the various sectors of the economy, attempting only to slow the increase of fossil-fuel combustion rather than embarking on major programs to develop alternatives for the *supply* of energy.

Such policies do not look beyond the short/medium term when the limits of energy-efficiency savings will be overtaken by the increased demands for energy. This will come with economic growth and development, especially when the expected swelling in demand in the developing world is taken into account (Anderson, 1995). It is with this *longer-term* perspective that this book is concerned: assessing the prospects for *mass* substitutions of fossil fuels by energy sources that do not threaten our environment and climate: sources that are inexhaustible and that provide secure access for the critical requirements for an energy input to the economy. Such sources will here be termed *sustainable* energy sources, including the solar, wind, and biomass technologies that have the potential to displace fossil fuels on a mass scale in the long term.

If there are alternative technologies for the supply of energy and there are ways of using energy more efficiently, it may be asked why these alternatives are not used if they can eliminate some or all of the problems created by fossil fuels. The answers, to date, are technological, economic, and otherwise behavioral and have been dealt with in many different treatises in recent years. This book deals with the prospects for the technological change in the energy industry that is needed for alternative sources of energy to substitute for the fossil fuels on a mass scale.

The purpose here is not to dwell on the market failures, failures of political leadership, or lack of industrial innovations for alternative energy technologies. These failures have made up the history of the period, starting with the energy crisis of the 1970s. Concerns during this period were focused, on and off, on the supplies of fossil fuel, principally oil. The level of anxiety rose and fell with world oil prices, and technological innovation was cut off when the price of oil fell. This, of course, was not a new pattern in American history and had been occurring since petroleum came on the scene a century ago (Cassedy & Grossman, 1990, 1998).

While not reviewing the history of the failures of alternative energy technologies, we will nonetheless be interested in the causes of failures in the past, insofar as they related to the innovation process itself. In the process of innovation, leading from invention to market adoption of new technologies, there are critical points where progress can stop. These failures in the process are rarely a consequence of the technology alone and frequently are entirely apart from the technical promise of the invention or new process itself. The most common obstacle, as mentioned above, has been the market, where a drop in competitive prices undercuts the anticipated profitability of the new innovation before it can be brought to the commercial stage.

There are other factors for the success (or failure) for the process; in some cases economics is the explanation, while some are explained by other theories of human behavior, such as sociology (Rogers & Shoemaker, 1971). The economic theories of innovations deal with factors of industrial firms, such as size, that are thought to influence the success of the processes of R&D (Mansfield, 1968). The sociological theories of innovations, by comparison, deal more broadly with the adoption of new technologies as they are involved in cultural change. We will explore both perspectives as we consider the various options that appear open to society.

The heightened concern over the environment during the 1990s has transformed our perspectives of technologies in general, with energy production, in particular, coming under intense scrutiny (Hollander, 1990). The possibility of global climate change has become a public issue for the longer term, while acute pollution conditions, as in the Los Angeles air basin, have galvanized urgent efforts for the near term. The emergence of the electric vehicle (EV), not considered a serious automotive competitor since the advent of the petroleum-fueled, internal combustion engine (ICE), is attributable almost entirely to the awakening to environmental realities for the people of California. As we shall see in this text, if the EV had continued to be judged on the basis of its present technical performance and costs, it still would not compete nor would it have had any likely prospects for the near future.

This approach to assessment

A major purpose for this book is to reassess the prospects for these alternative energy technologies in light of the growing public sentiment for an environmental approach and out of older concerns, such as energy security, which could return at any time. For the most part, the assessment will be of the renewable energy *sources*, such as those using solar, biomass, or wind resources. However, technologies that deal with energy end use, for instance those that store energy, such as batteries, will also be assessed. In some cases, these seemingly auxiliary technologies are crucial to the ultimate success of a source technology, with solar power plants or wind generators as prime examples of technologies that are reliant on auxillary support to be really effective. The EV is another example where storage not the primary creation of energy is crucial to success. However, we will not generally deal with entire energy systems, such as the entire energy conversion process in solar–thermal electric generation or the complete automotive-drive functions of an EV. Rather we will examine the attributes of these prospective technologies as basic sources of the energy required to perform their functions in the economy.

The prime criterion for selection of each prospective alternative for assessment will be that it could fit into an overall scheme leading toward sustainability overall for the supply of energy and the preservation of the environment. *Sustainability* is a term, or even a concept, that has been widely used in recent years, especially regarding international development. Brundtland (1987), for the World Commission on Environment & Development, has defined *sustainable development* as: "development that meets the needs of the present without compromising the ability of future generations to meet their own needs". This definition is the outcome of decades of debate worldwide over economic growth and the environment (Beder, 1993). It attempts to create an approach to the economic development demanded so vigorously, especially in the Third World, ". . . in which the material needs of all the world's people are met in ways that preserve the biosphere" (Brundtland, 1987).

In its broadest interpretation, it is taken to mean not only sustainability of resources and the environment but also social sustainability, including all aspects of economics, community, and social fabric (Brown, *et al.*, 1991; NRC, 1992; Daly & Townsend, 1993). It seeks ". . . an economy that exists in equilibrium with the earth's resources and its natural ecosystem . . ." (Allenby & Richards, 1994). Here, we will not be so ambitious as to encompass the broad definition but will only comment on salient cases of social sustainability, such as comparing centralized energy facilities with decentralized energy production. We will, however, be guided by criteria for energy technologies that are supportive of the broad concepts of sustainability (Steen, 1994).

In our assessment of energy technologies, that are in harmony with sustainability, we will be looking for one or more features of (i) inexhaustibility, (ii) renewability, or (iii) recyclability. By inexhaustibility, we mean superabundance, such as found with the solar technologies. Renewability is the familiar concept associated with hydropower but also is exemplified by biomass technologies. In the industrialized world (OECD) today, renewables (hydro, geothermal, and wood) account for only about 8% of primary energy consumption. Finally, recyclability can imply not only a renewing of the source of energy but also a recycling of undesirable products of energy conversion, such as the greenhouse gas carbon dioxide. Examples of this dual recycling concept would be managed energy crops or forests, where not only is the biomass fuel renewed but also the combustion product carbon dioxide is recombined in photosynthesis. Consequently, throughout our assessments of the various prospects, we will be looking for the presence or promise of these features of sustainability for an energy economy in the longer term.

The ideal of a near perfectly cyclable energy system, however, may not be technologically achievable, even in the long term. An example of such a system might be one of the "solar–hydrogen" proposals where solar energy is the primary source to separate hydrogen from water and the hydrogen is then used as an energy transmission medium. Ideally, of course, combustion or fuel-cell conversion of the hydrogen is pollutant free, having only water as the product of combustion or conversion. But such an ideal is achieved only with combustion using pure oxygen or conversion in some of the advanced fuel cells. The more practical and less costly use of air for combustion of hydrogen inevitably results in nitrogen oxide emissions, just as any of the fossil fuels do. The same is true of the much advocated use of natural gas as a means to reduce CO_2 emissions per unit of energy delivered and eliminate most sulfur dioxide emissions. Even though it is a less polluting fuel, natural gas combustion with air still results in nitrogen oxides emissions, in addition to CO_2.

Emissions are not, of course, the only possible impacts of candidate technologies to be considered for sustainability. Waste products and residues of the energy-conversion process should be scrutinized also, particularly with regard for any long-term implications. The most salient example of such is high-level nuclear waste, which presents health and security dangers extending into the future far beyond our society's control. It is for this reason, principally, that nuclear fission technologies (using uranium isotopes) have been excluded from consideration here. Nuclear fusion (using isotopes of hydrogen), where the nuclear wastes are not as long lived and do not pose long-term security, has been discussed but is questionable with regard to these sustainability criteria even if it were found to be technically feasible. The residue products of mining for alternative fossil fuels, such as oil shales and tar sands, also can have long-lasting environmental impact on more localized scales and, therefore, is questionable for fitting our criteria here.

In this book, we attempt a comprehensive assessment of the technological options open to society to supply its energy, looking forward to a goal of sustainability in the *long term*. The assessments will represent the author's best independent judgments of the realistic prospects, technically, economically, and environmentally. The emphasis will be on those technologies that, in the author's view, have the prospects for making major impacts on society's energy supplies well into the 21st century. In each case, the factors favoring sustainability will be evaluated in light of the current stage of development of the candidate technology. We will not give assessments to any extent on approaches, such as energy efficiency, demand-side management, or emissions reduction, that have been advocated to serve the transition in the medium term to a totally changed energy economy in which these sustainable technologies would be used widely.

Projections of the expanded use of renewable energy sources into the next century vary wildly. Starting from a base of total US use in 1990 of 6.8 QUAD (1 QUAD = 10^{15} BTU = 7.2 EJ), supplying about 8% of the nation's energy, these extrapolations range from 15 to almost 60 QUAD (16 to 63 EJ) by the year 2030 (Rosen & Glasser, 1992). (The 1990 base of use comprises mostly hydro and geothermal electricity plus wood burning.) The high estimates tend to come from organizations that advocate the use of renewable technologies, such as the Solar Energy Research Institute (now named the National Renewable Energy Laboratory) or the Union of Concerned Scientists and Public Citizens, whereas the more modest estimates originate from the US Department of Energy (DOE) and the Gas Research Institute, neither of which had dominant interests in renewables when these projections were made. A major objective of this book is to give insight into such forecasts, attempting to do this without underlying biases.

Critical assessment

These assessments will be of a critical nature, avoiding the promotional projections so common in reporting R&D progress. The prospects for gains or breakthroughs in technical performance or cost reductions will be cited wherever appropriate, but the optimistic projections of promoters will not be quoted. The uncertainty of the outcomes of R&D will be evident throughout all of our reviews of these new alternative technologies, especially with regard to the timescale for successes. The reader should not use the critical nature of the reviews to draw general conclusions that R&D projects on these ideas are unworthy of support – most of them are worth supporting. The author's intention here is merely to inject a note of skepticism on projections of R&D achievements by those having a stake in their future, no matter how carefully they may be constructed. The underlying motivation for this approach is to help to instill more realism into analysis and debate regarding future prospects. More discussion of such behavioral aspects will be made in the section on the R&D establishment and for specific examples in the assessments themselves.

This book is not written exclusively for the technologist, rather it should be understandable and useful to readers having a modicum of technological literacy on energy matters. Tutorial treatments for the non-specialist may be found in the science, technology, and society literature (cf. Cassedy & Grossman, 1990, 1998). The assessments are directed to business people and economists, as well as technical people, who are making business decisions or are formulating government energy policies. Students in graduate programs in business or public policy are also a prime readership for this book. In every assessment, an attempt has been made to bring out those technical features that are key to the success of the technology, without going into excessive detail. Any reader interested in further technical details will find them in the references, which have been cited at every point relevant to the assessments throughout the text.

The major objective here is to stimulate a critical appraisal by the reader of what the prospects are for each of these candidate technologies for a sustainable energy future, without the biases of promoters, and to be able to formulate a more realistic picture of where these technologies might fit into systems and schemes to meet the needs and contingencies of the future.

The presentation here will be discursive, avoiding the use of equations on technical issues, but at the same time being quantitative wherever essential for critical assessment. Figures are used mainly to convey a sense of working reality or to illustrate simple concepts, not to delve into technical detail. Economic competitiveness will always be quantified in terms of unit costs of energy that are readily compared with the market and competing technologies. This perspective is a market view, directed at broad business or public policy decision making rather than the financial analysis of individual investments in these new technologies.

Unit costs refer to the cost of energy produced per unit of energy supplied. For thermal energy we will use $/MBTU (US dollars per million BTU) or the nearly equivalent $/GJ (US dollars per gigajoule). For electric energy we will use $/kW-hr or cent(¢)/kW-hr. These unit costs will always be given on a basis adjusted for the depreciation lifetime of the equipment, following the "cost of energy" definition that is commonly used in the energy industries (see Appendix A). Unit-cost comparisons will generally be made with current market prices for conventional fuels and electricity, which appears justified in most cases during the present era of relatively stable energy prices. However, the prospects for changing energy market prices, both up and down, are discussed in several cases. Finally, in a few cases where it is appropriate for investment decisions, we will cite life-cycle cost comparisons.

The organization of this book

The book is organized into three parts: following Part I, the *Introduction*, Part II covers the *Candidate technology assessment* and Part III covers the *Prospects for technological change toward sustainability*. Part II includes nine chapters assessing the technological, economic, and environmental status of prospective, sustainable technologies. At the end of the book, the appendices cover *Energy cost analysis.*, a *Glossary* of terms and units used in the book, and *The conduct and management of research and development*. Major attention is given to solar energy sources, biomass energy, wind power, hydroelectric power, and energy storage because these technologies seem to have the best prospects for making major impacts on the energy economy in the long run. These topics are discussed in Chapters 1–5. Following these reviews are four chapters on geothermal power, ocean energy sources, nuclear fusion and hydrogen fuel from renewable resources. These later four chapters cover prospects that, in the author's judgment, present lesser promise for major impact within the foreseeable future. Every technology seems to have its promoters, but these appear to fall into a second ranking for one reason or another, which will be indicated. Part III summarizes all the prospects and discusses R&D within the field.

Not included in these assessments are the technologies appropriate for a *transition* to a sustainable energy economy (Flavin, 1992). These include ways to increase uses of fuels, such as natural gas, that emit fewer pollutants and greenhouse gases and to achieve efficiencies in the production or end use of energy, thus reducing these emissions from coal and oil. Cogeneration of electricity, which can combine a lower effluent fuel (natural gas rather than coal) with enhanced conversion efficiencies, is a prime example in electricity production. Heat pumps and higher efficiency appliances are good examples of increased efficiencies in end use. Increasing automobile gas mileage and the use of lower-emission gasoline blends are important examples in the transportation sector. All of these technologies are either presently in use (albeit to limited extents in some cases) or appear capable of development in the near to medium term. Several of them rely on the increased use of natural gas, supplies of which may become scarce a few decades into the 21st century, thus limiting their use only to the medium term.

The presentations in the assessment chapters will have common elements. Each chapter will start with background on the technical working aspects of the technology and the nature of the resource. These descriptions will not be extensive or highly detailed, always focusing on features crucial to feasibility. The current status, both with regard to technical operability and economic competitiveness will be reviewed, using available quantitative data. The regional availability and regional cost effectiveness will be covered wherever relevant. An important feature, not widely discussed elsewhere in the energy literature, will be the economics of *energy storage*. The status of ongoing R&D will be reported, including recent progress and results from the prototype projects that have been operated. These status reports will be confined to results already achieved and will not usually quote future projections on technical performance or economic competitiveness made by project managers or promoters of the technology.

Rather than use the projections of those with a stake in the outcome, the prospects for gains in technical performance or reductions in cost will be enumerated in each assessment as best as they can be judged independently. Comparisons will be made with competing technologies, whether existing conventional ones or other new alternatives. Wherever possible, plausible quantitative bounds will be cited for future technical measures and cost reductions, but seldom will there be any time scale put on attainment of such possibilities.

In any projection of market readiness of a new product or technology, there will always be the uncertainty of competitive price levels that has been the scourge of innovations in the energy field historically. It is pointed out in some cases of marginally competitive costs discussed here that policy instruments, such as price guarantees or, even, price supports (e.g. through tax credits), may be the only way to maintain the possibility of market entry at a critical juncture in time. The invocation of such instruments could occur if public policy were to demand them on any one of several grounds: protecting the environment, mitigating climate change, security of energy supplies, or looking to the ultimate exhaustion of fossil fuels.

References

Ager-Hanssen, H. (1993). *Energy for Tomorrow's World – The Realities, Real Options and the Agenda for Achievement*. World Energy Council. London: St Martin's Press.

Allenby, B.R. & Richards, D.J. (eds.) (1994). *The Greening of Industrial Ecosystems*. National Academy of Engineering. Washington, DC: National Academy Press.

Amirahmadi, H. (1995). World oil and geopolitics to the year 2010. *Journal of Energy and Development*, 21, 85–122.

Anderson, D. (1995). Energy efficiency and the economists. *Annual Review of Energy and the Environment*, 20, 495–511.

Beder, S. (1993). *The Nature of Sustainable Development*. Newham, Australia: Scribe Publications.

Bodansky, D. (1991). Overview of the energy problem: changes since 1973. In: *The Energy Sourcebook*, ed. Howe R. & Fainberg, A. New York: American Institute of Physics.

Brown, L.R., Flavin, C. & Postel, S. (1991). *Saving the Planet – How to Shape an Environmentally Sustainable Global Economy*. Washington, DC: World Watch Institute.

Brundtland, G.H. (1987). *Our Common Future. The World Commission on Environment & Development*. Oxford, UK: Oxford University Press.

Cassedy, E.S. & Grossman, P.Z. (1990, 1998). *Introduction to Energy – Resources, Technology and Society*, 1st & 2nd edns. Cambridge, UK: Cambridge University Press.

Cassedy, E.S. & Meier, P.M. (1988). Planning for Electric Power in Developing Countries in the Face of Change. *Energy Policy Studies*, 4, 53–100.

Chartier, R., Ferrero, G.L., Henius, U.M., Hultberg, S., Sachau, J. & Wiinblad, M. (eds.) (1996). Market, economic and environmental aspects of bioenergy, biomass – for energy and the environment. *Proceedings of the 9th European Bioenergy Conference*, Pergamon, Section 2.6. London: Elsevier Science.

Cleveland, C. & Kaufman, R. (1997). *Energy Journal*, 18, 89–108.

Daly, H.E. & Townsend, K.N. (eds.) (1993). *Valuing the Earth – Economics, Ecology, Ethics.* Cambridge, MA: MIT Press.

Ferrier, J.P. (Exec. Director, IEA) (1996). The world energy outlook: implications for economies in transition. *IAEE Newsletter*, 5, 8–11. Ferrier, J. P. (Exec. Director, IEA (1997). Globalization: challenges and opportunities in shaping a common future. *IAEE Newsletter*, 6, 4–6.

Flavin, C. (1992). *Building a Bridge to Sustainable Energy,* Ch. 3, *State of the World – 1992.* Washington, DC: World Watch Institute.

Goldemberg, J. (1996). *Energy, Environment and Development.* London: Earthscan.

Goldstein, L.J. (1997). Risks to global crude oil flow sustain need for strategic reserve. *Oil & Gas Journal*, Sept. 15, 20–24.

Hollander, J.M. (ed.) (1990). *The Energy – Environment Connection.* Washington, DC: Island Press.

IEA (International Energy Agency) (1995). *World Energy Outlook.* Paris: OECD.

IEA (International Energy Agency) (1996). *World Energy Outlook.* Paris: OECD.

Laherrere, J. & Perrodon, A. (1997). *Petrole et Techniques*, No. 406.

Mansfield, E. (1968). *The Economics of Technological Change.* New York: Norton.

NAS (National Academy of Sciences/Engineering/Medicine) (1991). *Policy Implications of Greenhouse Warming.* Washington, DC: National Academy Press.

NRC (National Research Council) (1992). *Global Environmental Change – Understanding Human Dimensions*, ed. Stearn, P.C. Washington, DC: National Academy Press.

New York Times (1997). *Half-Hearted. Global Warming Conference Closes Gloomily.* p. A10; *Clinton Defers Curbs on Gases Heating Globe*, p. A1, A11. *New York Times*, June 27.

Oil & Gas Journal (1996). Issue on world review of reserves. *Oil & Gas Journal,* 30 Dec.

Rogers, E.M. & Shoemaker, F.F. (1971). *Communication of Innovation: a Cross-Cultural Approach*, 2nd edn. New York: Free Press.

Rosen, L. & Glasser, R. (eds.) (1992). *Climate Change and Energy Policy.* New York: American Institute of Physics.

Salameh, M. (1996). Is a third oil crisis looming before the end of the 1990s? *IAEE Newsletter*, 5, 20–21, 26–27.

Steen, N. (ed.) (1994). *Sustainable Development and the Energy Industries – Implementation and the Energy Industries.* London: Royal Institute of International Affairs by Earthscan.

Weisel, J.H. & Kelly, J.E. (1991). US. Energy and the impact of acid rain legislation. *Journal of Energy and Development*, XVII, 99–120.

Further reading

Blair, J. (1978). *The Control of Oil.* New York: Random House.

EUROSTAT (Statistical Office of the European Community) (1995). *Renewable Energy Statistics 1989–91.* Luxembourg: ECSC-EC-EAEC.

Flavin, C. & Tunali, O. (1996). *Climate of Hope – New Strategies for Stabilizing the World's Atmosphere, World Watch Paper 130.* Washington, DC: Worldwatch Institute.

Lee, T.H., Ball, B.C. & Tabor, R.D. (1990). *Energy Aftermath*, Ch. 2, *Changes in the Energy Situation 1945–1989.* Boston, MA: Harvard Business School Press.

II Candidate technology assessment

1 Solar energy sources

1.1 Introduction

Solar radiation as a source of energy is, of course, the epitome of the clean, sustainable energy technology. Except for residues possibly arising out of the manufacture of solar components (e.g. semiconductors), solar technologies have very low environmental impacts. The environmental impacts of solar systems in operation are very low and the source is, for us, inexhaustible.

The energy incident on the earth from the wide electromagnetic spectrum emitted by the sun may be converted to useful heat, to electricity, or used to create a fuel. The uses of converted solar heat range from domestic hot water (DHW) to industrial process heat (IPH). Electricity may be generated from solar radiation either by thermal-plant methods, using solar-heated steam, or by direct conversion to (d.c.) electricity in solar cells. Alternative fuels, such as clean-burning hydrogen, can be evolved from solar-driven chemical reactions or by electrolysis driven by solar cells. Finally, of course, biomass fuels derive their energy from the sun. Any or all of these means of converting sunlight to useful energy could supply the worlds needs, if the technologies were ready. The problem with solar technologies is not, as is sometimes asserted, that there is insufficient land area to collect all of the energy society needs; worldwide there is and in the larger countries there is. In the USA, for example, it would take less than 2% of the land area to supply *all* of the country's primary energy demand from solar sources, at current consumption levels . The problem with solar energy, as we will outline here, is its *cost*.

The various possible ways in which solar energy can be used, replacing conventional fossil-fuel energy, are at different stages in technological development. In this chapter, we will assess the status of each of five prospective solar-conversion technologies, leaving biomass fuels and other indirect schemes to other chapters. The five technologies are solar DHW (SDHW), solar–thermal steam (for electric generation or IPH), solar (active) space heating, passive solar space heating, and direct solar–electric conversion (using photovoltaic (PV) cells). The first three of these involve the use of solar-heat collectors. Solar-heat collectors actually involve several different techniques, depending on the application, ranging from flat-plate collectors heating water to focusing collectors generating steam. By contrast, passive solar heating (and cooling) is a matter of architectural design. Finally, no heat collection whatsoever is necessary in the direct conversion of sunlight to electricity by semiconductor solar cells.

All of these attempts to achieve useful solar energy conversion have suffered from a lack of economic competitiveness of these new technologies and from the inherent defect of intermittency of the resource itself. Both of these deficiencies have prospects for solution in each of the approaches being studied. Technological innovations have attempted to lower construction costs of solar collectors and to provide energy storage against intermittency. Most of the five technologies appear to be evolving, albeit fitfully, toward eventual success in the markets for energy. There is still potential for new concepts to revolutionize the prospects in regard to both costs and intermittency. In any case, progress will likely only be attained in the long run with sustained support of R&D efforts, whether from direct public support or through public policies that encourage private invesment. This should become clear in the following sections as the history of these efforts is revealed.

1.2   Solar domestic hot water

Solar collectors on the roofs of homes are the most common image (Fig. 1.1) of solar technology. Such collectors have been used, for the most part, for SDHW systems to supply hot water for the house or to warm water for swimming pools. There have been over half a million such systems installed in the USA since the 1970s.

The majority of SDHW systems are termed *active* systems, meaning that it takes a circulating pump, operated from sensors and a controller unit, to make them function (Fig. 1.2). The coolant fluid must be circulated through tubes in the collector (see Fig. 1.3) and carried down to the heat exchanger in the hot-water tank in order to transfer the collected solar heat into the water. The controller causes the pump to circulate the coolant fluid from the collector through the system only when the sensors detect a temperature difference sufficient to transfer heat into the hot-water storage tank. Some designs, called "thermosiphon", operate on the natural convection of the temperature difference without the need for a pump, but these still require an active control system to function.

The technology of solar hot-water systems is relatively simple. Apart from the collector, the system is essentially plumbing and controls. The flat-plate collector (Fig. 1.3) should have surfaces that maximize the absorption of sunlight and minimize the reradiation of heat. Absorption is easily obtained simply by the use of a dull black coating. Lowering reradiation, however, is attained only with the use of more sophisticated and, therefore, more expensive coatings. The collector must also have a transparent cover to contain the heat once it is collected. Provision must be made in cold climates against freezing of the circulating fluid, such as the use of antifreeze solutions for the coolant or a drainback provision if water is the coolant. A heat exchanger coil transfers heat from the circulating coolant into the (potable) water in the hot-water tank in most systems.

A typical home in the sunbelt region requires 100–150 ft^2 (9–14 m^2) collector area to supply most of its hot water needs, but as little as 60 ft^2 (6 m^2) can supply 50% of needs in some sunbelt locations (Hof, 1993).

Figure 1.1
A solar domestic hot
water (SDHW)
collector. (Courtesy of
the American Solar
Energy Society.)

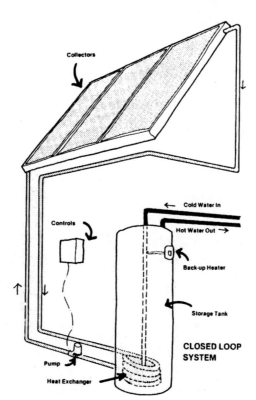

Collectors

Cold Water In

Hot Water Out

Controls

Back-up Heater

Storage Tank

**CLOSED LOOP
SYSTEM**

Pump

Heat Exchanger

Figure 1.2
A solar domestic hot
water system. (*Source*:
Northeast Sustainable
Energy Association.)

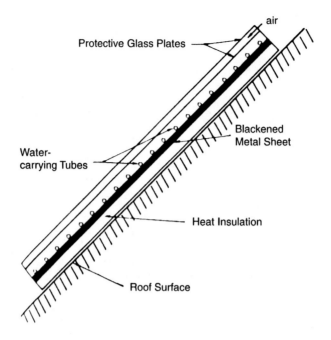

Figure 1.3
A flat-plate solar
collector.
(*Source*: Penner &
Icerman, *Energy*, vol. II,
© 1975,
Addison-Wesley
Publishing Co., Inc.,
Reading, MA.
Reprinted with
permission.)

Household hot water typically runs 50–100 gallons daily, requiring around 50 000 BTU (53 000 GJ) heat. An auxiliary (back-up) heater is required in most sections of the country to meet the needs during cloudy periods. Generally, it is uneconomical to attempt to store hot-water heat to cover such periods and SDHW designs look for the least-cost "solar fraction" trade off between the solar system and its back-up (Beckman *et al.*, 1977). This fraction, based on annual operation, typically runs 60–90% in the sunbelt and 40–60% elsewhere in the lower 48 states of the USA. Similar comparisons hold for northern versus southern (sunbelt) regions elsewhere. Hot-water storage equivalent to about 1 day's usage is recommended both to carry over the evening and early-morning needs and to smooth supply on days of intermittent sunshine.

If storage is attempted to achieve a high solar fraction approaching totally-solar operation, hot-water storage capacities equivalent to 2–3 days of use will be required in sunbelt sections and more in other regions. Increases in storage capacities above this level yield ever diminishing returns. An attempted totally solar system (i.e., no back-up heater), besides being uneconomical, will most probably leave the household with insufficient hot water for some weather sequences. The larger the storage and collection capacity, the smaller the probability of insufficiency, but it is impractical to try for total coverage. An account of possible modes of solar operation with storage is given in Winter, *et al.* (1991). A more detailed discussion of the economics of hot-water storage is given here in Chapter 5.

The challenge for solar manufacturers has been to fabricate systems that are reliable and durable, yet cheap enough to compete against conventional (fossil-fueled) systems. For example, multilayer wavelength-selective coatings or (low heat loss) evacuated tubes to improve the efficiencies of collectors have generally resulted in costs too high for the DHW market. Correspondingly, it is prohibitively expensive for SDHW to have the collectors track the sun – collectors are simply oriented south facing (in the northern hemisphere) with a fixed elevation tilt that optimizes annual collection over the seasons. Collector frames are typically made of inexpensive extruded aluminum and insulated with cheaper forms of fiberglass or plastic foams. At the same time, however, performance requirements, such as collection efficiency, low heat loss, and durability against high temperatures and ultraviolet radiation, must compete against the cost factors for an overall product competitiveness.

The history of SDHW shows wide variations in its use (Hof, 1993). Solar hot-water systems, using simple roof-mounted tanks, were installed in Southern California early in the 20th century, but their use declined after 1920. Florida residents also utilized solar hot water, with some 50 000 units being sold in the decades 1920–50. Use declined in both sunbelt states with the advent of cheap oil for heating. Even wider use of solar hot-water collectors was made in Japan and Australia prior to the 1970s but also declined with the availability of low-priced fuel oil. With the arrival of the oil crisis of the 1970s, interest revived worldwide in various alternatives such as solar heat.

With the revival of SDHW in the 1970s, its path to adoption in the market has been checkered in the USA (Frankel, 1986; Larson & West, 1996). Low-quality products and fly-by-night marketers, in the midst of an industry otherwise trying to gain experience with the peculiar requirements of the new technology, created a wariness in the buying public (FSEC, 1979). Attempts by the fledgling industry to establish standards and warranties, which would have put DSHW systems on an acceptance basis comparable to that of conventional boilers and hot-water tanks, were slow in the making. Such uncertainties for the solar consumer, together with the inevitable barrier of the newness of the technology, made for few adoptions in the potential market even without considering the costs (Vories & Strong, 1988). Even though testing procedures and standards (SRRC, 1993–4; Larson & West, 1996) have subsequently been developed and industry certification now exists for SDHW systems, the opportunity of the 1970s was largely lost.

The principal determinant for the low market penetration of DSHW systems, however, has been price. Even when oil and other energy prices were at their highest in the 1970s, exceeding (wholesale) price levels of $5/MBTU, solar hot water was competitive only in the sunbelt. In general, it took incentives in the form of investment tax credits to promote sales of DSHW systems. A growing industry, selling annually tens of millions of square feet (collector area, a measure of system size) of systems, existed into the 1980s when the federal tax credit had been set at 40% (Andrejko, 1989) and states, such as California and New York, had added credits of their own up to 55% for domestic systems and up to 25% for commercial buildings. Perhaps no better indication of the necessity of the incentives was the collapse of the DSHW market after the expiration of the federal tax credits in 1985. Up to that point, in the previous decade, over 100 million square feet of collectors (9 million square meters), with systems, had been sold benefiting from the subsidies. While this failure was deplored by solar advocates and many others, the 40% measure of price competitiveness, at least for the sunbelt market, was a demonstrated indication of the market status of the technology in the absence of aid to early adopters.

Progress has taken place in the technology since the mid-1980s, some changes improving operational efficiencies and durability, and several innovations have led to reduced fabrication and installation costs (Andrejko, 1989). Operational modes, involving slower circulation of the coolant and improved heat exchanges, have improved system efficiencies. The use of improved absorbing coatings and foam insulation has improved collector efficiencies by 15–20%. The lighter weights of the insulation and use of aluminum frames have resulted in easier and less costly installation. New materials, such as unbreakable (tempered) glass on collectors and more durable absorbing coatings, have extended the working lives of collectors. New "brushless" electric motors for circulation pumps have improved system performance and reliability. Finally, the innovative use of photovoltaic cells as combined solar sensors and electricity suppliers for the controllers of the circulation systems has resulted in reduced cost and in operational enhancements.

Technical progress alone, however, is not likely to be the key to market penetration for DSHW. Technical improvements are likely to continue incrementally only, given the relatively "low-tech" nature of these systems (Katzman, 1984). It continues to be apparent that the major determinant of market competitiveness of DSHW heat will be its price compared with those of conventional fuels (oil and natural gas). That price comparison for the DHW market can be made on a unit energy basis, such as $/MBTU. The unit cost of the SDHW energy is determined strictly by the pay off of the capital investment over the life-time energy output of the equipment, plus (small) operating costs.

As an illustration, consider a hot-water system in the sunbelt, where there is 680 000 BTU/ft^2 (7722 MJ/m^2) of annual insolation (250 W/m^2, average), a 50% collector efficiency, a system unit cost of $50/ft^2 ($538/m^2) collector, and a 25-year depreciation lifetime. This system would yield unit energy costs of $17 per MBTU ($16.1/GJ), including 10% interest charges and small operating costs levelized over the life of the system (see Appendix A for the cost of energy methodology). This cost might be compared with $9/MBTU for heating oil at $1.25 per gallon or $7/MBTU ($6.60/GJ) for natural gas at 70¢ per therm (0.1 MBTU), which a home owner in the USA might be faced with in the 1990s. Clearly, this is not a competitive cost of energy in the market for the commodity of heat energy with the current level of energy prices.

The average homeowner, however, is more likely to look at the fuel savings on the solar hot water to pay back the initial purchase price of the system and not explicitly consider interest costs (Cassedy & Grossman, 1990, 1998). In this simplified view, the homeowner considers every solar-produced BTU as a saving on a BTU produced by the auxiliary (back-up) fuel-burning heater. By this simplified view of life-cycle costs (see Appendix A), the homeowner could calculate that 340 000 BTU/ft^2 (3860 MJ/m^2) annually collected solar heat (i.e. sunbelt insolation, with a 50% collector efficiency) would pay off the investment in 16 years, if oil is the alternative fuel, or 20 years, if the alternative is natural gas. If interest was included, it would take over 35 years to show a life-cycle payback, and then only for a very low interest rate. A consumer attitude survey (Stobaugh & Yergin, 1979) has found that a payback time less than 5 years is commonly necessary to attract serious consideration of purchase by the public. In a few cases, utility customers have been lured into attractively financed purchases of SDHW systems, subsidized by the utility as part of their demand-side management (DSM) programs (Carlisle & Christiansen, 1993), but otherwise no financing schemes have been devised that would be attractive to the public when the basic cost of energy is in the range that includes solar domestic heat.

The inescapable conclusion is that, in the absence of financial incentives and with prevailing fossil-fuel prices, domestic solar hot-water systems are not economic, even in the heart of the sunbelt. In other parts of the country, the annual solar production would be half or less that of the sunbelt and the unit energy costs or payback times twice or more. This situation could only be changed by a jump in fuel prices over twice the present levels or by tax incentives of more than 50% for the sunbelt (more elsewhere), or some combination of the two. The first possibility has a historic precedent in the 1970s but is very ulikely to be repeated in the first part of the 21st century. Second, the prospects of government policy in the USA returning to subsidizing solar commercialization are problematic in the absence of another world oil crisis or other widely accepted contingency, such as slowing climate change. In other OECD countries, the prospects for solar domestic price supports might be somewhat better. Subsidization of a new technology is likely to be accepted as sound policy in some of these countries only if that technology appears to need just temporary assistance in becoming competitive in the market. In the author's estimation, with small prospects for dramatic reductions in SDHW costs through technical breakthroughs, this lack of cost competitiveness does not seem to be temporary.

The only other argument that could be mustered in support of temporary tax subsidies would be that mass production would lower present unit investment costs of DSHW systems to a fraction of the present \$50/ft^2 level. While such arguments seem plausible for other technologies, such as photovoltaics, they do not seem appropriate for these comparatively rudimentary systems, where dramatic technological advances seem unlikely. If that is the case, then there would have to be a rationale for subsidization of an indefinite duration, such as being part of a broad policy of reduction of fossil-fuel use. The current lack of interest, at the political level, in energy policy and the general turn to *laissez faire* politics make any such policy quite unlikely.

As of 1991, there were about 1.25 million SDHW systems installed in homes in the USA, representing only about 1% of its market potential (Golob & Bus, 1993), and these mostly as a result of the tax credit program of 1977–85. In Israel, 65% market penetration exists, partly because of the insolation level giving high solar-fraction operation and partly because SDHW systems are required by the government in all new homes and (small) apartment buildings (Shea, 1988). This regulation is based on a national policy to reduce dependency on imported oil. Much the same motivation exists in Japan, where over 5 million solar hot-water heaters are already installed and more are installed each year. The sense of urgency for the USA to adopt a national policy to reduce oil imports has long since passed. It is possible, however, that environmental policy, such as that enacted in California against air pollution, could become the rationale for solar subsidies.

1.3   Solar (active) space heating and cooling

Active space heating

Space heating using solar collectors and *active* transfer of the heat is the same concept as (active) solar hot water but on a larger scale. It takes considerably more collected energy, and a proportionally larger collector (Fig. 1.4), to keep an entire house heated on a winter's day than it does to supply just the household hot-water needs. As a result, the commitment by a home owner to a solar space-heating system is much larger than that for only hot water. For example, requirements run up to 100 ft^2 (9.3 m^2) collector per room to be heated, with each square foot of collector costing in the vicinity of $50 (system cost) (Hof, 1993). Solar space-heating systems have also been installed in commercial buildings, with very similar design and cost considerations as domestic systems but on a larger scale.

Solar space-heating systems can be based on either hot water or hot air. The hot-water installations are larger versions of SDHW circulating systems, delivering their heat to radiators just as conventional fuel-fired furnaces do. Hot-air systems (Fig. 1.5) by comparison, circulate air through a solar collector (with air ducts) and into the rooms of the home or building. Whereas some heat storage had been thought necessary in space-heating systems to smooth the heat supply, some newer hot-air, space-heating systems do not have it. These systems operate on ambient fresh air (Andrejko, 1989) and are found satisfactory for commercial buildings in daytime use only. Where heat storage is necessary for hot-air systems, large porous beds of pebbles through which the air is forced by fans serve as inexpensive storage media.

Where heat storage – either hot water or hot air – is necessary to maintain round-the-clock space heating, it must be used in conjunction with auxiliary heating for an economic system in temperate climates. Attempts to increase the solar fraction for space-heating systems by increasing storage capacity and/or collector area is even more a matter of diminishing returns (Beckman et al., 1977; Hof, 1993) for space-heating

Figure 1.4
A domestic space heating collector. (Courtesy of the National Renewable Energy Laboratory, Boulder, CO.)

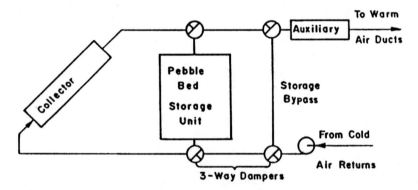

Figure 1.5
Hot-air space heating system. (Courtesy of the US Department of Energy.)

systems compared with hot-water systems. (This may be recognized as simply a consequence of the seasonal mismatch of solar availability with space-heating needs.) Nonetheless, a totally solar-heated house has been constructed and operated in the sunbelt, although it was not deemed optimum on a cost basis. Optimum (annual) solar fractions appear to be in the range of 75% in the sunbelt and down to the 40–50% range in northern climates of the U.S.

The recent history of active solar space heating parallels that of SDHW. During the energy-crisis period, with tax credits, there were over 25000 systems sold (Andrejko, 1989). For the most part, these were retrofits of existing homes. The manufacturers and distributors were generally from the same group as suppliers of SDHW and suffered from the same deficiencies of standards and quality (Hof, 1993). Federally sponsored demonstration programs were introduced early in the period (mid-1970s) but did little to assure industry standards and reliability of the equipment sold when the tax-credit program was initiated. Later, the industry was able to improve on quality and institute a certification system for collectors. Nonetheless, the public image of solar systems, space heating as well as hot water, suffered. Furthermore, sales collapsed, as they did for solar hot water, with the end of tax credits for solar home use in 1985.

Active solar cooling

Active solar cooling is a seemingly self-contradictory term that is used in space-cooling technology, such as absorptive refrigeration, which can use collected solar heat rather than fuel-fired heat to drive a thermodynamic cooling cycle (Duffie & Beckman, 1974; Hof, 1993). In such operation, the solar collector system supplies the absorptive refrigerator with the hot working fluid, which is the energy input needed to drive the cooling cycle. The principle of operation of solar-driven absorptive cooling is the same as that when natural gas is used to supply the heat for the thermodynamic cycle.

A solar cooling system, for summer use, can be combined with a solar, space-heating system for winter use. Flat-plate collectors, of the type described earlier, can be used if the cooling cycle uses a lower temperature coolant, such as lithium bromide. The joint-purpose system (Fig. 1.6) can then be adjusted for either season's space-conditioning needs by use of the right-hand valve shown in Fig. 1.6 (the valve is used to bypass the solar subsystem only). For winter's heat, the valve is set to let the hot water flow through the heating coil, while the valve set for summer cooling allows flow through the (absorption) air-conditioning unit. Room air is forced through the joint unit for either operation. The unit can also supply domestic hot water the year round.

The joint use makes for a better payback on the investment in both the collector and its subsidary equipment, with a much improved capacity-factor utilization of the solar source. Summer solar cooling, in particular, is a good match of the load demand to the source, in contrast to that of winter solar heating. However, the principle barrier to market adoption has been, as with solar in general, the initial cost. The current installation costs for solar cooling, with solar-system costs allocated just to cooling, are three to eight times those of conventional cooling systems (Andrejko, 1989). Even with solar-system costs allocated to both cooling and heating, however, there would be no savings on energy costs over conventional fuels, such as natural gas, for the technology in its present state.

Not surprisingly, solar cooling had a history in the "energy crisis" period paralleling that of solar heating and hot water. The sales of solar cooling units grew with the help of tax credits but dropped abruptly with their demise in 1985. In the 5 years the credits were given, nearly 10000 solar cooling units were sold. This was after government-sponsored proto-types had demonstrated operating feasibility. At the end of the 1990s commercial heat-driven cooling systems are marketed with natural gas burners as the heat source, not solar collectors.

It is worth noting, however, that the potential market for solar cooling is

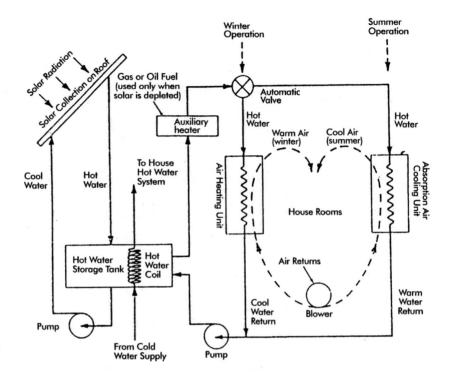

Figure 1.6
A solar heating/cooling system. (Courtesy of G. Lof, 1973.)

huge (many billions of dollars annually), if one is to view the entire air-conditioning energy demands. These are estimated to be 2.3 QUAD annually in the USA (about 3% of all end use) and comparable amounts in developed countries located in temperate and tropical climates elsewhere. Since air-conditioning is required in regions having high levels of insolation in the summer season, it is reasonable to consider this entire demand as a potential market matching the solar source to the cooling needs. This technologically optimal situation in warm regions will also be an economic advantage when solar cooling approaches market competitiveness.

In conclusion, we see that active solar heating and cooling is in the same situation as solar hot water in its prospects for widespread adoption, in temperate climates especially. It seems doubtful that technological gains in solar heat collection for either purpose can overcome the low price of fossil fuels in the world economy at the end of the 20th century. While dual use (heating and cooling) may give a competitive edge in sunny regions if solar collector systems become marginally competitive, the prospects seem dependent more on the future of conventional energy prices than technological innovations in low-temperature solar collector systems.

1.4 Passive solar heating and cooling

Passive solar utilization does not depend critically on the development of a new technology for its success, it is largely a matter of architecture and construction design by methods well known to professionals in the field

Figure 1.7
Passive solar architecture used in a home in Colorado. (Courtesy of the National Renewable Energy Laboratory, Boulder, CO.)

(Cook, 1989; Anderson, 1990). In the winter season, the objective is to collect and retain solar heat to the maximum extent possible, purely by building design rather than active operation (Figs. 1.7 and 1.8a). This is accomplished by use of large window areas (south facing), good heat insulation, and large heat-storage masses for retaining the absorbed sunlight, such as brick walls (Fig. 1.8b).

For new construction, passive solar heating can save up to 70% on conventional heating in sunbelt locations and even 30% in "gloombelt" locations in the USA and Europe (Thayer, 1994). Added costs will be incurred for the solar exposure and retention features, but the payback periods will be short as a result of the significant fuel savings possible with good design. Retrofit construction costs, however, will vary widely and paybacks would have to be evaluated case by case for net benefits. The development of new materials, such as insulation or solar absorbents, has the potential to improve efficiencies or reduce construction costs (Babomf, 1992; Balcomb, 1992; Wright, 1992; Haggald & McMillan, 1993; Viceps, 1993) and some government programs function to assist the building industry to make such improvements (Larson & West, 1996). These improvements, however, can be expected to be incremental in nature accompanying refinements in design and not occurring as a result of major breakthroughs in prospects.

Passive cooling, unlike passive heating, is less a matter of construction and design than site selection, house orientation, and the landscaping accompanying new construction (Anderson, 1990). Window exposures north and south are important as is siting within the shade of existing trees and added shrubbery for shade. In the structural design, large roof overhangs (Fig. 1.8) work well to give summer shading but still allow winter warming when the sun is low in the sky (Miller, 1998). Clerestories with openable windows allow natural venting of warm air. Providing for natural ventilation through screened windows and doors is another simple addition to passive cooling design, which can be supplemented with the use

(a)

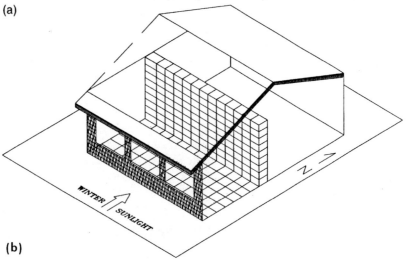

Figure 1.8
Passive solar design
(a) A passive solar
home in North Carolina.
(b) A passive solar
home design
(Photograph courtesy
of the National
Renewable Energy
Laboratory, Boulder,
CO.)

(b)

of fans (e.g., ceiling fans). Finally, the use of light-colored roofing, radiant barriers under the roof, and radiant reflecting glazing on windows, for summer exposure, add significantly to the passive cooling design. Again, no innovations in technology are required for such measures to be effective for energy conservation and providing comfort.

The history of the adoption of passive solar designs in the USA has been noticeably different from the active solar technologies. Government programs for passive solar development have been carried out more along the lines of support of commercialization, rather than direct subsidy of investment (Frankel, 1986). Thus, information dissemination, training, and consumer surveys were typical of these demonstration and commercialization activities for as long as they were supported in the early 1980s. Even though cost savings should have been evident to prospective home buyers, adoptions were not as large as might have been expected. Initial investment costs were clearly a barrier to adoption, but so also was awareness of the economics of energy consumption. All told, about 200 000 new homes and 15 000 new commercial buildings were built in the USA with solar architecture during the decade at the end of the energy crisis in the 1970s. Again, with the subsequent drop in concern with falling oil prices, interest fell here as it did for the active solar technologies during that era. Some interest has developed in the 1990s in coupling passive solar construction of new homes with utility DSM programs (Aitken & Bony, 1993).

Significant savings in fuel costs could still be had even at the level of fuel prices of the mid-1990s (markedly lower than the oil-crisis period). A major barrier to further adoption of passive solar designs is a defect in the marketing of new home construction, which emphasizes the initial price of a new home rather than the overall life-cycle costs. This situation could be altered to some degree, even in the present energy market, through government dissemination of information to the public, at little public cost. Whatever gains are to be made, however, will be extremely slow because of the very low rate of turnover in housing.

In summary, we see that the prospects for widespread adoption of passive solar heating and cooling are more in the realm of commercialization and market penetration than in technological innovation. The progress of adoption will depend on both public policies and energy markets, principally oil. Much will depend on the markets for new housing and retrofit construction, together with public awareness of the life-cycle savings offered by passive solar investments. These prospects are, therefore, uncertain and depend on factors that are qualitatively different to those involved in other solar technologies. The outcomes in adoption will not turn on technological developments but on markets, public policies, and other non-technological factors.

1.5   Solar-thermal technologies

Solar–thermal technology is the term that has been applied in the solar R&D community to solar applications for the generation of industrial heat and thermal generation of electricity, which are under development (Winter *et al.*, 1991; Larson & West, 1996). For the most part, they involve the use of solar concentrators (e.g. dish or trough reflectors) which result in higher operating temperatures for active, solar collector systems. Many of the prototype solar-thermal systems generate steam, either for direct use in industrial processing or to drive steam turbines in otherwise conventional thermal–electric generation schemes.

Solar concentrators

Solar concentrators are able to focus the sun's (direct) rays ideally into a perfect point or, alternatively, along a line (Winter *et al.*, 1991). In fact the laws of optics say that the perfect point is really a small spot and the ideal line is really a linear strip of finite width. It is parabolic-dish reflectors (Fig. 1.9a) that focus the sunlight into a spot and parabolic trough reflectors (Fig. 1.10) that focus the sun along a narrow strip. Another type of solar concentrator which collects the sun's rays into one small area is the "central receiver", using a large array of mirrors (Fig. 1.11a). Each of these mirrors is oriented to reflect the solar flux into a central-receiver volume mounted on a tower (Fig. 1.11b).

With both the dish reflector and the central receiver, the concentrated (ratio 1000:1 or more) sunlight is absorbed into a receiving surface inside a central volume, thus generating temperatures in the range 500°–1500°C (900–2700°F) (Winter *et al.*, 1991). As such, this volume is a "solar furnace" and may be used either to generate steam for electric-power generation or for high-temperature, low-impurity materials' processing. In some prototype central receivers, high-temperature heat-transfer fluids such as molten salts have been used instead of steam/water. For either type of point focus, volumetric collector systems, the sun must be tracked in both azimuth and elevation through its daily trajectory.

Tracking with dish collectors (Fig. 1.9) is accomplished by simply orienting the reflector to be normal to the incoming solar rays, using mechanisms similar to those used for tracking-radar antennae. For the central receivers, each of the reflecting mirrors (called heliostats) is oriented individually in order to track the sun along its daily trajectory (Fig. 1.11c). In the case of the large Solar One project (Andrejko, 1989), this means over 1800 heliostats (Fig. 1.11a) have to be controlled individually.

A parabolic trough collector focuses the solar flux along the axis of a tube (Fig. 1.10) through which the heat-transfer fluid passes. The tubes are coated with a solar absorbent layer and the heat is transferred to the fluid (hot oil or pressurized water) by conduction. The concentration ratios of trough reflectors (in the range 100:1) are lower than the point-focus types and consequently the temperatures reached are not as great (80–400°C or 150–750°F) (Duffie & Beckman, 1974; Beckman *et al.*, 1977; Winter *et al.*, 1991). Applications of the hot water or low-temperature steam from trough collectors include IPH and steam-driven electric generation. Individual line-type collectors are interconnected so that the (heat-transfer) water passes from one collector to the next (Fig. 1.10) in a large field of such collectors (Fig. 1.12). The modularity of the units allows for additions to the fields merely by adding to the rows of collectors.

Trough collectors must also track the sun but can do so in a vertical angle only. A common mode of tracking with troughs is with the collector axis aligned along an east–west line, following the sun's elevation as it first rises to its apogee and then falls back to the horizon. Another way uses a north–south collector axis, tracking the plane of the sun in its daily path from east to west. This method favors summer-time collection when the sun rises highest in the sky. Neither method of single-axis tracing can be as effective as the two-angle tracking of point-focus concentrators, however, especially over the seasonal variations of the sun's diurnal path.

Although interest in focusing the sun's rays for heating dates back to

(a)

Figure 1.9
Parabolic dish solar collectors. (a) Solar furnace concentrator at Sandia National Laboratories. (Courtesy of the National Renewable Energy Laboratory, Boulder, CO.) (b) Solar engine dish at Science International Corporation. (Courtesy of the American Solar Energy Association. (*Photo credit*: David Patryas.)

antiquity and the principles of optics are well known in modern physics, engineering projects to utilize high-temperature solar heat have emerged only in the 20th century. Some interest sprang up around the turn of the century from Europeans attempting to use solar dishes and troughs to generate steam (Dunn, 1986). This was more in the nature of a scientific curiosity than a serious attempt to develop a new technology, however, in view of the ever increasing availability of fossil fuels at that time. Similarly, solar-thermal projects were not started in the USA until the oil crises of the 1970s. These projects have been carried out, with varying levels of support, under federal sponsorship or under private initiative with government incentives (Gupta, 1987; DOE, 1992). A thorough history of the US solar-thermal program, complete with analysis of the conduct of its various phases, has been given by Larson & West (1996).

US federal R&D programs have emphasized the point-focus techniques

(b)

(DOE, 1993; Holl & de Meo, 1990), with the central receiver having the most sustained support. Experimental central receiver plants have also been built in Europe, Israel, Japan, and countries of the former Soviet Union (Gupta, 1987; Brower, 1992; de Laquil, *et al.*, 1993), sponsored by the EC and by individual governments. The major prototypes in the USA, and the world's largest to date, have been the Solar One and Solar Two solar-tower projects in Barstow, California (Fig. 1.11a,b), which demonstrated the operating feasibility of the central receiver concept. Solar One was built in the late 1970s under US DOE sponsorship and operated successfully for 5 years, supplying electricity to the grid of the Southern California Edison Co. (SCE) with a generating capacity of 10 MW_e. The plant operated with a superheated steam system (Figure 1.11c), with an input temperature of 510 °C (950 °F) and utilized digital controls for its 1818 heliostats.

Figure 1.10
Parabolic trough solar
collectors. SEGS
troughs at Kramer
Junction, CA.
(Courtesy of the
American Solar Energy
Association.)

Solar One was shut down in the late 1980s, having served the purposes of the project at that time, and some of the facilities were later used in a cosmic ray study. No further prototypes were initiated during the 1980s since the DOE policy of that era confined support to development of new solar components with no new commercial prototype projects started. In 1991, a cost-sharing effort between SCE and DOE was initiated to improve and expand the project, which will have a 100 MW$_e$ capacity and be called Solar Two (Vant-Hull, 1992).

Several technical improvements have been incorporated into the design of the new central receiver, including its heat-transfer fluid (liquid sodium), heat-storage medium (molten salt) and heliostats (silver-laminated acrylic membranes) (Vant-Hull, 1992; de Laquil, 1993). In addition, smoother thermal operation will be achieved by delivering the collected solar heat directly to storage and drawing the steam heat from storage (by heat exchanger) for the steam turbines. This will smooth the cyclic, high-temperature operation of the system and ease many mechanical/plumbing problems. New high-temperature working fluids, such as molten salts, are also being investigated for these alternative operating modes.

Many of the design innovations planned for Solar Two have had improved plant economics as the objectives and these improvements should lead to still others. It should be possible to increase the thermal efficiency of the steam system beyond the 31% achieved with Solar One, thus approaching that of conventional power plants operating with super-heated steam. Also, in the Solar Two design, the use of coated-membrane mirrors was part of an attempt to reduce the cost of the heliostat field below the unit-area cost ($400/m^2 or $37/ft^2) of Solar One (Winter *et al.*, 1991), which used glass–metal mirrors. The US DOE has calculated the levelized cost of electricity from central receivers scheduled for demonstration in 1995 to be in the range $0.10–0.20/kW-hr (Trieb, 1997). Finally, the

(a)

(b)

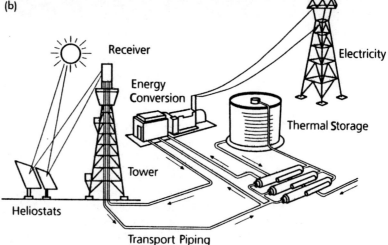

Figure 1.11
Central receivers.
(a) Solar Two Project.
(Courtesy of the
American Solar Energy
Association.) (b) Solar
One pilot plant with its
central receivers at the
top of a tower.
(Reprinted with
permission from
Johansson, T.B. *et al.*
(eds.) (1993).
*Renewable Energy –
Sources for Fuels and
Electricity*, Island Press,
Washington, DC and
Covelo, CA.) (c) Solar
central receiver
water-steam system.
(Adapted from Boer,
1990.)

(c)

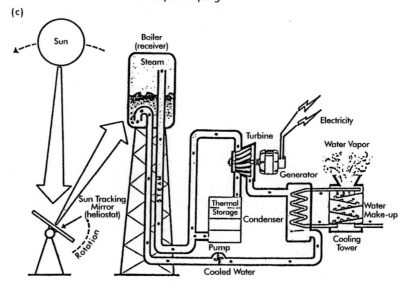

Figure 1.12
Ariel photograph of LUZ SEGS Plant, Kramer Junction, CA. (Courtesy of the American Solar Energy Association.)

use of liquid sodium as the heat-transfer fluid allowed a higher temperature input than the previous (oil) coolant, thereby improving the thermal efficiency. The resulting increased electric output would again help to improve the payback on investment.

Smoothed receiving operation on Solar Two using storage would serve to increase the long-term output of the plant, also improving the capacity factor and, therefore, the pay back on investment. The annual capacity factor (CF) (see Appendix B and Marsh, 1980) of Solar One, using no storage, was 38%. It should be recognized, however, that it was operated in peaking or intermediate load service for the SCE utility and, therefore, CFs comparable to conventional base-load plants could not fairly be expected. If storage heat were to be used, this particular plant could meet the late-day demand of the SCE load profile by period-displacement operation (see Chapter 5), thus increasing its cost savings for the system.

Prototype demonstration projects using line-focus trough collectors have been carried out in the USA, Spain, and Japan (Trieb, 1997). The largest of these was in the USA and was sponsored, either wholly or partially, by the DOE in the late 1970s. These projects supplied IPH in the temperature range 85–420 °C (150–750 °F) to over 15 firms engaged in industrial processing, such as pharmaceutical, brewing, leather tanning, chemicals, and enhanced oil recovery (Brown, 1983). In the lower part of this temperature range, water/steam was the heat-transfer fluid used in the trough collector tubes, whereas special oils were used in the higher range. While initial design and operational defects in the new technology hampered some projects, operational success was achieved for the most part, even though economic competitiveness was marginal at best.

The energy policies of the 1980s terminated federal demonstration projects for trough concentrators, as well as for other solar technologies mentioned here, and new initiatives in solar industrial process heat nearly ceased. A few projects, including a brewery and a commercial laundry, were built in the early 1980s (Brower, 1992) and were able to benefit from a continuation of federal solar energy tax credits (Becker, 1992). These projects were able to deliver process heat at a levelized cost of about $6.15/MBTU ($5.83/GJ), which was not quite competitive with the low heating costs of natural gas ($4–5/MBTU) at the time (near the end of the energy crisis period).

Decision making by industrial or commercial potential users of solar heat has been influenced by the payback they would get for their investment. Having or being able to enlist financial expertise, they could conduct full payback analyses to aid their decision making. Accordingly, complete life-cycle or discounted cash-flow analyses could be carried out. In these analyses, they would account for the fuel savings over the life of the solar investment, discounted back to the time of investment or levelized over that lifetime. This would be dependent, of course, on the particular situation of the industrial user, especially regarding regional fuel costs and regional annual insolation. The projection of future fuel costs is an area of uncertainty; while this is not as much a concern in the 1990s, it could become a major issue again.

Returning to the history of solar-thermal development, a combination of incentives in the 1980s, particularly in California (Lotken & Kearney, 1991), led to the most successful US project to date in solar-thermal technologies. The application of the trough-concentrator technology to steam-turbine electric generation enabled the LUZ International firm to combine federal and state (California) solar tax credits to yield a total of 38.5% net price reduction in cost of energy to their customers. They also utilized the federal PURPA (*Public Utilities Regulatory Policy Act* of 1978) and State of California market guarantees for independent power producers, thereby creating a financial footing that was viable for a few years. A total of nine plants were built and successfully operated over the period 1984 to 1991. Three more projects were in planning when the termination of the federal credits in 1991 set off a chain of events leading to the bankruptcy of LUZ. The operation of some of these projects has subsequently been taken over by Constellation Energy, Inc. of Baltimore, Maryland (Williams & Bateman, 1995).

The nine different trough-concentrator plants in Southern California built by LUZ International were called the Solar Electric Generator Stations (SEGS I (1984) through SEGS IX (1990)). These started with a 14 MW$_e$ plant producing electricity at a levelized cost of $0.24/kW-hr (Becker, 1992), followed by a series of six 30 MW$_e$ plants and ended with two 80 MW$_e$ plants, generating at costs in the range $0.08–0.12/kW-hr (Carlisle & Christiansen, 1993; de Laquil *et al.*, 1993). About 85% of these costs are from the initial capital investment, divided almost evenly between the collectors, thermal–electric conversion, and the balance of plant. Subsequent (1993) reports from projects of other types of solar-thermal electric generation have indicated operating costs in the range $0.09–0.13/kW-hr (DOE, 1993; Ahmed, 1994).

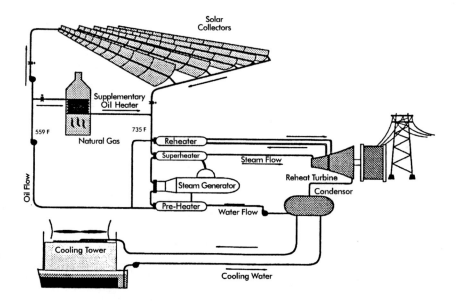

Figure 1.13
The LUZ SEGS system.
(Courtesy of the
American Solar Energy
Association.)

The later SEGS plants operated with plant thermal efficiencies of nearly 38%, an increase of over 6% over the first plants. Technical improvements were made over the series of plants, including increased mirror reflectivities (over 90%), reduced vulnerability to wind damage, fewer receiver tube fractures, and increased coolant outlet temperatures (over 400 °C or 752 °F). Operational improvements in washing the trough mirrors helped to maintain reflectivity in the dusty desert climate. The plant output over time was well matched to the daily load–demand profile of SCE Co. For peak loads, the $0.08/kW-hr was competitive with the cost of production by the generating plants otherwise available to SCE during those periods.

The last of the SEGS series of thermal solar power plants were 80 MW$_e$ capacity each, all located in or around Harper Lake, California (Lotken & Kearney, 1991). For this output power, they required large fields of trough collector assemblies (852 of them) each assembly being 100 m long (Figs. 1.12 and 1.13). The heat-transfer fluid was pumped through 142 parallel loops, with each loop containing six assemblies arranged in series. The system, in full sunlight, could raise the heat-transfer fluid (oil) to 410 °C (770 °F) as it passed out of the collectors into the steam generator and heaters (heat exchangers) (Figure 1.13). Using this oil system, superheated steam could be generated, which is most efficient and appropriate for steam turbines.

It will be noticed (Fig. 1.13) that a supplementary heater, burning natural gas, is included in the SEGS system. This heater provides back-up either for cloudy periods or to extend the daily output into the evening to meet the electric utility's load–demand time profile. When in use, the gas heater also increases the thermal efficiency of the plant by raising the input temperature to the turbine. LUZ has been limited by PURPA rules to a maximum of 25% natural gas (Becker, 1992). For the most part, however, the solar energy input is well matched in time, during a sunny day, to the (air conditioning and business) load demands of SCE.

There is a modest R&D program ongoing for development of solar-dish electric generation (Gupta, 1987; Brower, 1992) in the USA and in other countries. The USA funded solar thermal R&D for almost $24M in 1995 (Larson & West, 1996) (see Chapter 11 for overall funding levels). Much of this work has emphasized the development of new materials and components, such as thin-membrane dish reflectors and a high-temperature, sodium-vapor, solar-absorbing and heat-transfer medium. The use of higher temperatures is part of efforts to raise thermal efficiencies of solar-thermal systems in general. The experiments to date, however, give no indication of commercial operability as experienced with the Solar One central receiver or the SEGS series of trough concentrators.

No trial markets have been developed for parabolic-dish concentrators (Brower, 1992) as they had for the central receivers and trough concentrators. Successful experiments have been carried out using parabolic dishes for steam production of electricity (Gupta, 1987; de Laquil *et al.*, 1993). The largest prototype installation had an array of 700 individual dishes with a combined output capacity of nearly 5 MW$_e$ (Holl & de Meo, 1990). Most of these prototypes have been in the kilowatt range, however. Some of these smaller units have been operated with a thermal input temperature of 800 °C into a Stirling engine as the prime mover for the electric generator. Concentration ratios have ranged from a few hundred up to 8000. Thermal efficiencies in some of these solar dish experiments have reached 30%. In summary, it would seem that dish concentrators have been consigned to small modular units of 100 kW$_e$ or less and that applications would be limited accordingly to stand-alone operation or possibly a grid-support role.

Progress in solar-thermal concentrator technologies has been achieved since the start of their development in the 1970s, despite the uneven support accorded them. The Federal Energy Bill of 1992 reinstated energy production tax credits for renewable energy sources and provided for cost sharing up to 50% for most renewable-energy projects (Westby, 1994), although this credit was not renewed by Congress in 1998. In the political climate of the 1990s, it is quite unlikely that such support would continue in the USA. Whatever national policies will be, the general requirements for commercial viability of solar-thermal sources should include the following interdependent goals in order to approach market penetration:

- collector area-dependent costs less than $150/m^2
- thermal-energy production costs less than $5/GJ ($4.75/MBTU)
- electric-generation capacity costs less than $1500/kW$_e$
- component (long-term) durability
- high system availability and low maintenance.

Operations meeting these production requirements require sunbelt insolations (2500 kW-hr/m^2 or 230 kW-hr/ft^2 thermal, annually). While none of these technical/economic requirements has yet been met by any operating prototype, significant progress has been made toward them.

So far, those solar-thermal concentrator systems that have been successfully operated have been of moderate output capacity: less than 400 MW_{th} or 100 MW_e, which is about one fifth of the capacity of a new conventional, fossil-fueled "base-load" plant. Whereas it is a straightforward matter of system design to make modular increases in the capacities of line-focus (trough-concentrator) systems, point-focus systems (central receivers or dish concentrators) do not lend themselves to modular expansion. It seems likely, under present conditions, that point-focus generating systems will be limited to a few hundred megawatts of electric output capacity for any single installation as a result of the inherent flux density and temperature limitations of the receiver volumes. The largest system contemplated to date is a central receiver of output capacity 200 MW_e, which would require a land area of 10 km^2 to accommodate over 12 000 heliostats (de Laquil, et al., 1993).

The lack of technologies for large-scale, economical energy storage (either thermal or electrical) relegates these large solar-thermal installations (e.g. Solar One and SEGS) to providing intermittent service only (see Chapter 5). For delivery of electricity to a utility grid, this means only displacing high-cost conventional generation (typically, combustion turbines) and then only when peak demand has some coincidence with peak solar generation. This will, in most cases, limit the usable solar-thermal capacity to about 15% of the installed capacity of the utility grid (see Chapter 5 for further discussion). The same would apply to central PV fields feeding the utility grid. (SEGS extended this utility service to "intermediate" load durations by using the natural-gas back-up heaters.) When operated for solar industrial heat, back-up heaters are normally provided in order that the industrial processing is not interrupted, since thermal storage cannot supply more than a few hours of the heat energy needed.

Solar ponds and solar bowls

A common goal (the reduction of capital costs), achieved by a common approach (using an excavated cavity in the earth) unite these two otherwise dissimilar solar-collection techniques. The solar pond, on the one hand, is a means to collect and store unconcentrated sunlight. The solar bowl, on the other hand, uses the cavity as a fixed dish-type reflector, which concentrates the solar rays onto a movable receiving element for tracking.

Solar ponds
The most common type of solar pond contains concentrated salt water, either naturally occurring or created artificially. They operate from a temperature inversion induced by maintaining a gradient of salt density in the pond that increases with depth (Fig. 1.14). If the density of salt near the bottom of the pond is higher than at the surface, then this bottom water is heavier (higher specific gravity) and will not rise (convect) when heated.

(a)

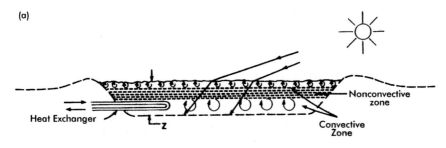

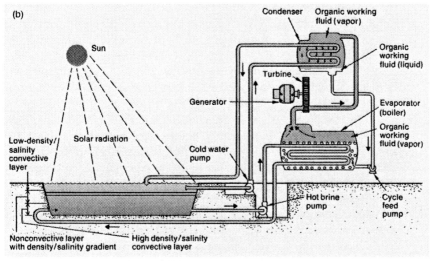

Figure 1.14
A solar pond. (a) A cross-section of the pond. (Adapted from Winter, 1990.) (b) The heat collection system. (With permission from Bronciki, Y.L. (1981). A solar-pond power plant, *IEEE Spectrum*, Feb., 56–59, © IEEE.)

Consequently, as sunlight penetrates the pond, the water at the bottom can get heated to a temperature much higher than that at the surface without rising to the surface and losing its heat. By this method, the temperature near the pond's bottom can be raised to near the boiling point of water (100 °C).

Some salt-gradient ponds exist naturally (Zangrando & Johnson, 1985) in regions of the USA and elsewhere. Some artificially excavated ponds use natural brines located near the site. Sea water, evaporated in basins to increase solution, has also been used (*Solar Energy*, 1991). Others are entirely artificial, with their sites selected for high annual insolation and low latitudes (for more nearly vertical solar incidence). Artificial ponds of varying types have been constructed in Israel, India, Italy, Australia, Argentina, and Saudi Arabia, as well as the USA.

An alternative to the salt-gradient pond is the solar-gel pond . A layer of gel floating on the surface of a pond can block heat convection and insulate against the loss of solar heat that has been absorbed by the water below. In some early versions of gel ponds, some salt is used in the water to increase the buoyancy of the gel, but with the development of lighter gels it should be possible to use fresh water. The advantages of the gel pond are mainly operational: not having to maintain a gradient of salt concentration. The major disadvantage is cost, since the cost of the gels tends to add significantly to the expense of the projects.

The solar heat collected at the pond's bottom can be extracted by circulating the hot brine through a heat exchanger outside of the pond (Fig. 1.14b). The heat can be exchanged to make hot water, used for industrial processing or simply for space heating of buildings (Andrejko, 1989; Zangrando & Johnson, 1985). Since the solar heat is effectively stored at the bottom of the pond, its extraction in daily operation can be delayed until evening, for purposes such as space heating. Other heating uses include crop drying or industrial processing using air heated by the heat exchangers. Solar ponds are also used in mineral processing, either by direct heating of the process solutions or by heat exchanger of the hot brine into the process solution.

The other use of solar-pond heat is for thermal generation of electricity. In this application, the hot-brine heat exchanger delivers the solar heat to a working fluid, which in turn drives a turbine coupled to an electric generator (Fig 1.14b). Since solar-pond systems operate at temperatures much lower than ordinary steam-turbine systems, an organic liquid that has a high vapor pressure (at these lower temperatures) is used as the working fluid. While these solar-pond power plants have been demonstrated to operate well, the thermal efficiency cannot be expected to approach those of steam plants because of the Carnot (thermodynamic) limits of the operation at lower temperatures. Nonetheless, even with operating thermal conversion efficiencies less than half those of the higher temperatures using steam, the low cost of construction of these plants results in a competitive price for the electric energy generated.

The development of solar ponds was pioneered in Israel, starting in the 1950s (Bronicki, 1981; Lodhi, 1989). Several of these projects have been sited near the Dead Sea and use salts derived from those brackish waters. The largest demonstration plant to date is a pond 0.25 km^2 in area generating up to 5 MW_e and there are plans to quadruple that capacity. The busbar cost of electric energy from these prototype plants had reached the competitive range of $0.10/\text{kW-hr}$ by the late 1980s (Andrejko, 1989).

The promising economics of solar ponds has also been confirmed in the USA, accounting for all capital costs, interest, inflation, taxes (paid and credited), and operation and maintenance costs (Zangrando & Johnson, 1985). The low capital costs of ponds can be appreciated by considering the (collector) area-dependent costs, which at $75/\text{m}^2$ ($7/\text{ft}^2$) are one fifth or less of those achieved so far for concentrators, troughs, and dishes and one half of the commercial viability criterion ($150/\text{m}^2$ or $14/\text{ft}^2$) cited above for solar-thermal technologies in general. Still, these economics of solar ponds can only be achieved where the combination of high annual insolation, large land areas, brackish water, and salt supplies is available, for example in Israel or the American West.

The ability of the ponds to store solar heat, as well as collect it, has a particular advantage when electric generation with "period displacement" of energy for peak loads is the application (see Chapter 5). The thermal system for the ponds, transferring heat from the hot brine, is much faster than other thermal technologies using conventional boilers (Bronicki, 1981). Thus, a solar-pond plant has a start-up time of only a few minutes, comparable to the fast response time of combustion turbines, which are conventionally used as stand-by for utility peak load demands.

Some technical and environmental problems require further R&D efforts with solar-pond programs, as they are carried out in the USA, Israel, and other countries. A long-standing concern has been the maintenance of the required gradient in salt density over the depth of the pond (*Solar Energy*, 1991). Wind-driven waves tend to cause mixing of the solution near the surface, which becomes more of a problem as larger ponds are constructed with longer unobstructed distances for the wind to build up waves. Undesirable mixing also occurs deeper in the ponds where the brine is being extracted or re-injected. In addition, maintaining gradient layers deeper in the pond becomes more problematic with increasing pond size because of difficulties in handling larger bodies of water to maintain stable gradients. Unless solved, this could limit the maximum achievable capacities for solar-pond plants. Other technical issues include maintaining water clarity for solar penetration and reducing heat losses to the ground.

A major environmental concern for the technology is leakage of the brine from the pond into ground waters. An impervious liner is usually required for solar ponds. The detection and repair of leaks, without service outages, is an essential short-term requirement; the identification of synthetic materials for durable liners is a longer-range requirement.

The market potential for solar ponds is substantial, despite the regional restrictions discussed above (Zangrando & Johnson, 1985). For agricultural processing, over 0.75 QUAD (0.79 EJ) per year of primary energy could be substituted nationally for fossil fuels in the USA. For distillation processes in desalination of water, where the sites are appropriate for the ponds anyway, the US national potential is nearly 0.67 QUAD (0.71 EJ) per year. Finally, the potential for solar-pond electricity is estimated to be about 3.5 QUAD (3.7 EJ) per year by the turn of the century, restricted to sunbelt regions only. The realization of these potentials in commercial market penetration will require, as with the other solar technologies, the right combination of private investment, government-sponsored R&D, and policy incentives against market uncertainties.

Solar bowls

The solar-bowl technology is an attempt to reduce significantly the investment costs of concentrator, solar-thermal projects. Reports from an early prototype (Lodhi *et al.*, 1991) show (collector) area-dependent costs of approximately $50/m^2 ($4.60/ft^2), which is one third the commercial viability criterion that is cited in general for solar-thermal technologies. A better comparison is with solar ponds, since the bowls are subject to the same regional restrictions for performance. There we find the area-dependent initial cost of bowls are two thirds those of ponds. These reductions of initial costs, even at the expense of operational efficiency, are thought to be particularly important for projects in developing countries.

The solar-bowl construction consists of an excavated hemispherical bowl in the earth that is lined with a cement shell (Fig. 1.15). (Point-focus reflectors are *parabolic* "bowls"). Hundreds of small mirrors are attached

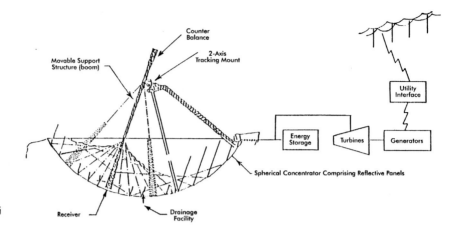

Figure 1.15
A solar bowl. (With
permission from Lodhi
et al., 1991.)

to the shell of the bowl, thus approximating a concave hemispherical
mirror. The spherical concentrator focuses the suns rays along a line
passing through the center of the sphere, in contrast to the action of a
parabolic concentrator, which focuses to a point. A line-type receiver, in
the form of a tube, is mounted above the bowl, with a two-axis pivot
located at the center of curvature of the hemisphere. The receiving rod
tracks the sun simply by being pointed towards it and all incoming solar
rays are then focused along the axis of the rod. Tracking by this system is
only effective at relatively high elevation angles, 45° either side of the
vertical. (The bowls themselves are not complete hemispheres.) The net
effect of this is a relatively short operational day for solar collection, even
in equatorial regions (Lodhi *et al.*, 1991).

The solar-bowl concept has so far been tested only with a few prototypes
(in Texas and Malaysia). Operation has been satisfactory, generating
superheated steam temperatures around 800 °C (1470 °F). A 20 m diameter bowl, focusing the sun onto a receiver tube of length 5.5 m delivered
250 kW_{th} thermal energy, enough to generate 100 kW_e electricity. Projects
up to 10 MW capacity are considered feasible. The technology is unique in
that it can deliver high-temperature steam with resulting high thermal-
conversion efficiencies, comparable with that of conventional power
plants, and yet is a simple, low-cost fabrication. As mentioned above, its
cost at this early stage of development appears to be a fraction of those of
the other solar–thermal concentrator technologies.

Solar bowls have excellent market potentials in developing countries
because of their simplicity and low maintenance, in addition to their low
initial costs. Applications include high-temperature industrial heat and
thermochemical processing, in addition to electricity generation. The technology would be ideal for economic development programs in that the
fabrication can be done using indigenous workers, who require only
technology transfer assistance and not the importing of "high-tech" equipment from the industrial countries. Finally, unlike other evolving solar
technologies, low costs have been achieved initially through the simplicity
of the basic design and, therefore, are not dependent on a subsequent

transition to mass production. Further development and prototype testing will be needed before broad marketability can be established.

Summary of solar thermal technologies

Prospects are limited for large-scale use of high-temperature solar heat as a substitute for fuels in generating electricity or supplying IPH in the medium-near future. Large solar concentrator systems have not yet reached the stage of development to be competitive with fossil fuels, even considering the marginally commercial operation of the expansive trough-collector fields in California. Their success could only have taken place in remote, sunbelt locations and with the benefit of tax credits. The largest of these trough-collector plants, although covering immense areas, were only 100 MW_e in output capacity, which is small compared with a larger conventional power plant (700–1200 MW_e). Even more restricted by region are solar ponds; this limits their use despite their advantage of lower cost over the concentrators. Solar bowls are promising as low-cost, technically simple systems for Third World use but are still in the development stage. Both solar ponds and solar bowls would be restricted to generating or industrial installations of modest capacity (a few megawatts) only.

It would appear that prospects are not good for decades to come for solar–thermal technologies to make a significant impact on the national energy economy. The LUZ trough prototypes were the closest to competitive for the cost of electricity (Trieb, 1997). Further technical developments are needed on the various forms of solar–thermal collection, some more than others. All will need further operational refinements and cost reductions if they are to be brought within the competitive range of cost per unit of energy produced (capital plus operational). Even then, low fossil-fuel prices, as long as they persist, are likely to undercut solar-thermal energy prices (unsubsidized). The situation in that sense is similar to that for the low-temperature technologies (e.g. SDHW), where technological advance may not be able to overtake the energy market within any foreseeable time.

1.6 Direct Solar Electricity

Direct conversion of sunlight to electricity is feasible without having to go through the intermediate stage of heat conversion (Maycock & Stirewalt, 1985). This was demonstrated on a working level in the space program of the 1950s using the photovoltaic effect in semiconductors (Green, 1982; Cook, 1991) and is currently the subject of research using photochemical reactions in liquids (Ohta, 1979). In semiconductors, photovoltaic conversion takes place in the narrow vicinity of a p–n junction between two regions of different impurity doping (Fig. 1.16a), when electron/hole pairs have been created by incident sunlight. The p–n junctions are also the heart of operation of modern day transistors. Solar cells are semiconducting slices or thin films having the required junction region doping, with

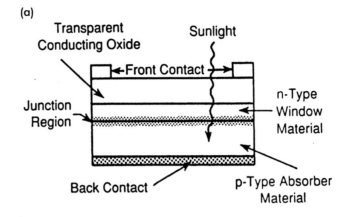

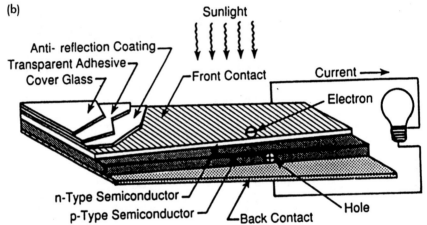

Figure 1.16
Photovoltaic cells. (a) Semiconductor junctions; (b) cell assembly. (From Howes & Fairberg (1991) reprinted with the permission of Springer-Verlag, New York.)

appropriate electrical contacts, and they deliver d.c. electricity (Fig. 1.16b).

The PV solar cells are today a working technology for many special applications (Strong, 1987; Hubbard, 1989) (Figs. 1.17 and 1.18), but are too costly for widespread use for the bulk supply of electricity. Nonetheless, limited adoption has proceeded for "remote-site" electricity supply, which is very useful where conventional power is not available. These applications include locations such as wilderness facilities, cattle ranges, or irrigation in the Third World. Similarly, PV installations have been found of great utility in rugged-site service, such as communication stations (Figs. 1.19 and 1.20).

Even in regions where conventional power is available, PV installations can be cost-effective alternatives to long power lines. In addition, a large commercial market has developed for consumer products such as hand-held calculators, which use PV cells supplying only minute amounts of electricity. In all of these applications, the technology is working with low maintenance requirements, and it is doing so reliably even under harsh physical conditions.

The major objective of R&D in PV technology since the 1980s has been

Figure 1.17
Solar panels in a roof top installation at the SMUD Solarport in Sacramento, CA.(Courtesy of the American Solar Energy Association.)

Figure 1.18
A remote site photovoltaic system supplying power to a rural home in Brazil. (Courtesy of the American Solar Energy Association.)

to reduce the cost of the cells (Stobough & Yergin, 1979; DOE, 1991), in order that the price of delivered PV electricity would be comparable to that of conventionally generated electric power (at the generating station bus-bar). Consequently, research programs have looked for ways to reduce semiconductor and fabrication costs, on the one hand, and for increased conversion efficiencies, on the other. Since a chief indicator of the cost competitiveness is the price of PV cells per unit of (peak) electric output ($/W_p$), these two objectives have often been in opposition to one another. Increased conversion efficiency increases the electric power output of the

Figure 1.19
A photovoltaic-powered microwave relay station in Argentina. (Courtesy of Siemans Solar Industries.)

cell for the same solar flux input, but costs more to achieve. However, lower-cost semiconductors and cheaper fabrication will usually lower the conversion efficiency. Also, cost-cutting techniques can result in compromised performance or durability of the cells.

Crystalline semiconductors, such as silicon ingots grown in a single crystal from the molten state, were the basis for the original PV cells and are still today the materials giving the best technical performance (Johansson *et al.*, 1993; Kazmersky, 1997) and contributing 85% of the market (McConnell *et al.*, 1998). (The basic understanding of semiconductor behavior is grounded in the regular lattice structure of a perfect single crystal (Green, 1982).) Single-crystal silicon cells have demonstrated stable conversion efficiencies (see Appendix B) exceeding 20% for 20-year, low-maintenance lifetimes of service. Recent enhanced designs may have efficiencies approaching the theoretical limit of 24% for single-junction, crystalline silicon cells. The cost of a single-crystal cell, however, is too high because of the costs of the ingot growing, slicing, doping, and polishing, and because of the unnecessary bulk of the expensive crystalline material itself. (Solar cells need only to be several optical wavelengths in thickness.)

Research strategies were established very early to look for thin sheets or films to reduce the bulk of semiconductor required (Carlson, 1989; Ullal & Zweifel, 1991). It was found that the bulk of costly material could be greatly reduced by using either thin sheets drawn from the molten state or thin films deposited from the gaseous state. Also, these forms lent themselves to possible mass production, with the promise of reductions in the unit costs of module fabrication. Another major thrust has been to try to move away from perfect crystalline semiconductors. Polycrystalline silicon, for example, had early been recognized (Green, 1982; Mitchell & Surek, 1988) as maintaining much of silicon's semiconducting properties,

Figure 1.20
A photovoltaic-powered
desert telemetry
installation.(Courtesy of
Siemans Solar
Industries.)

even though the material is made up of microscopic, randomly oriented
crystalline grains rather than a single perfect crystal. Using special film-
growing techniques (Hubbard, 1989; Johansson *et al.*, 1993), thin polycrystalline ribbons or sheets can be drawn from the molten silicon, using
far less silicon for a solar cell of given area than the ingot method. After
doping, the polycrystalline sheets or ribbons perform in solar cells nearly
as efficiently as single-crystal silicon does. Prototype polycrystalline cells
have been reported to have a conversion efficiency of 14% or more (DOE,

1995, Kazmersky, 1997). However, even though the polycrystalline sheets could be produced at about two thirds the cost of crystal silicon, the unit costs (in dollars per watt peak output) of polycrystalline cells were still not in the competitive range (by the mid-1980s) because of low conversion outputs.

A somewhat startling innovation has been the attempt to use amorphous (that is, materials having no crystalline structure) silicon. Thin films of amorphous silicon (a-Si) or other semiconductors can be deposited from vapors or hot plasmas onto supporting substrates (Cook, 1991; Johansson et al., 1993). The amorphous silicon, amazingly, still retains some of its semiconducting properties, although solar-conversion efficiencies have been lower than those of either crystalline or polycrystalline silicon. Considerable improvement occurs with hydrogenated silicon, where the films were deposited from vaporous silicon hydride. The hydrogen impurity is thought to restore more of the crystalline structure by filling defect sites in the interatomic linkages. The differently doped regions, furthermore, can be deposited one after the other in making the films, rather than have a separate doping process; this achieves a more cost-effective process.

While a-Si films show great promise (DOE, 1991), their conversion efficiencies are generally half or less those of crystalline cells after they suffer from output degradation upon initial exposure to full sun. Stabilized efficiencies have only been able to exceed 12% (Kazmersky, 1997). Even though these shortcomings pose barriers to widespread adoption for electric power conversion, a-Si cells presently comprise a major share of the market for PV cells in consumer-electronics products. Progress is being made, however, on improving the stability of a-Si cell efficiencies, and significant gains in efficiencies have been made with multijunctioning (Johansson et al., 1993; DOE,1995) (see below). Even before this, however, the current PV market is about 30% polycrystalline silicon (Williams & Bateman, 1995).

Semiconducting compounds such as gallium arsenide (GaAs), copper indium selenide (CuInSe), and cadmium teluride (CdTe) hold promise of higher efficiencies than silicon because their wavelength bands for sunlight absorption are more nearly optimum for the solar spectrum (Green, 1982; Boer, 1990, 1992). The band structure of CdTe is especially well matched to the terrestrial solar spectrum (Kazmersky, 1997), which has certain wavelengths deleted by absorption in the earths atmosphere. The cells can be fabricated readily from these compounds in the form of deposited thin films (Ullal & Zweifel, 1991; Johannson et al., 1993). Deposition of GaAs on substrates of similar crystalline structure yields crystalline thin films and these have demonstrated conversion efficiencies somewhat higher than those of crystalline silicon in a comparison of single-junction cells. Deposition can be done rapidly and with relatively little energy input using vapors, sputtering, spraying, or other means.

These compounds are not only more absorptive of sunlight than silicon but offer possibilities of multijunction cells for wide-spectrum conversion (see below). In addition, thin films of CuInTe and CdTe have been found not to degrade under initial sun exposure, which does occur with a-Si thin films. While these last two film compounds have suffered from lower efficiencies than silicon, not yet reaching 15% for single-junction cells, better results have been attained using the compound CuInGaSe, (over 16%) (Anon., 1996). Nonetheless, it should be recognized in making these assessments that the most promising use of these thin-film compounds will be for multi-junction cells where there are possibilities for higher efficiencies.

Multi-junction cells represent attempts to broaden the limited band of solar wavelengths that can be absorbed and converted photovoltaically in a single p–n junction. Semiconductors are able to convert light energy to electric energy only over a limited range of wavelengths, thus restricting the amount of energy from the full spectrum of incoming sunlight that can be utilized. This is a consequence of the inherent energy "band structure" of each semiconductor, which allows only certain ranges of photon energies to be absorbed for conversion in the p–n region. However, by stacking junctions of different semiconductors together, the various absorption wavelength bands of each can be included for conversion and the fraction of the incident solar energy available for conversion can be increased overall.

Thin-film, multi-junction cells have been fabricated using combinations of the arsenic compound and a-Si semiconductors (Hubbard, 1989; Boer, 1990, 1992). Conversion efficiencies for tandem single-crystal cells have been reported to be 10% or more than from single cells. Some further improvements in performance can be achieved optically with antireflection coatings and using solid-state tunneling regions within the cell to interconnect the tandem junctions. The technique will likely find its limit with tandem stackings of three cells, with conversion efficiencies reaching a little over 30%. In addition, the costs must be reduced and durability demonstrated for the multi-junction cells in order for them to be considered for commercialization.

The use of focusing reflectors or lenses (Fig. 1.21), similar to those used in solar-thermal designs, have been considered as a means of converting more solar energy in a PV cell (Boer, 1992; Johansson *et al.*, 1993) since the concentrator collects the solar flux from a wider area. In addition, the efficiencies of most cells will increase (up to a limit) as the solar intensity (watts per square meter) increases. Solar concentration should reduce unit capacity costs (dollars per peak watt produced) because more solar energy is available to the cell at the same cost. Focusing, however, has the added costs of tracking and only concentrates the direct rays of a clear sky, as is the case for solar-thermal designs. Even though tracking PV concentrators can result in a larger annual insolation collection, these factors run counter to the gains in unit costs. Tracking, for example, appears to add 50% to the "balance of system" (BOS) costs. Candidate cells for the concentrators are the semiconductor thin films mentioned above and it is hoped that mass production using these semiconductors will reduce unit costs of the cells themselves. Overall, further development and design work are needed before commercialization is considered.

Innovative features in the structure and packaging of PV cells have also

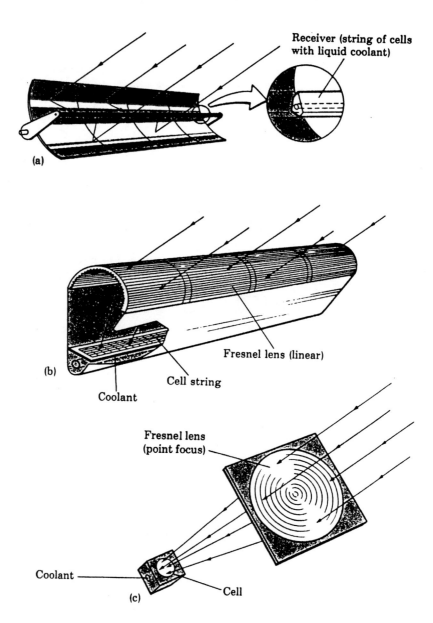

Figure 1.21
Methods for
concentrating sunlight
on photovoltaic cells.
(a) Parabolic trough; (b)
linear Fresnel lens
arrangement; (c)
point-focus Fresnel
lens.(Courtesy of the
US Department of
Energy.)

been tried by developers to improve performance. One idea is the use of
textured receiving surfaces and reflecting back surfaces (Fig. 1.22), which
results in a multiplicity of internally scattered light rays, called "trapping"
(Green, 1982; Cook, 1991). Another idea is the use of buried p–n junction
regions, which reduces light shadowing of the electrodes and achieves
lower semiconductor losses (Kazmersky, 1997). Finally, the use of low-
cost metallurgical silicon in the form of tiny beads on aluminum foil may
be an option for reduction of the fabrication costs of cells (Levine *et al.*,
1992).

Figure 1.22
A textured photovoltaic cell.
(Credit: M.A. Green *et al.*, Proceedings of the 19th Photovoltaic Specialists Conference, 1987; © IEEE 1987. cited in Cook (1991).)

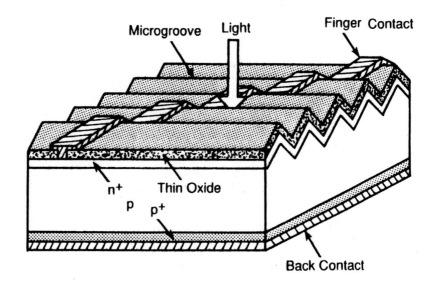

Most of the advances in photovoltaic technology just described have taken place over the two decades since the energy crisis of the 1970s. The R&D in PV technology has fared much better than that of the solar-heat programs in the US during this era. While federal funding was cut during the 1980s, there was still some continuity of support for PV programs. Under the National Photovoltaics Program of the DOE, an R&D partnership was built between government, industry, and university efforts (Mitchell & Surek, 1988; DOE, 1991, 1995). DOE funding for photovoltaics was at $73.5M for the fiscal year 1995, down from a peak of around $150M in the 1980s. R&D programs have also received the support of governments in Great Britain, Brazil, and Israel, and research is carried on in several European countries under the sponsorship of the Commission of the EC (van Overstraeten & Caratti, 1988; Johansson *et al.*, 1993) and the IEA (Morse, 1997).

Thin-film cells of various types have been fabricated with conversion efficiencies exceeding 20% without focusing and over 30% if concentrated by focusing (Ullal & Zweifel, 1991; Stolte, 1994; Kazmersky, 1997). Module efficiencies, accounting for shadowing by electrodes and encapsulation of the cells, are a few percent less. Theoretical limits on cell efficiency are around 30% unfocused and 40% focused, depending on the semiconductor and number of tandem junctions used. If such efficiencies could be achieved with long-time reliability, the mid-term technical goals of the DOE National Photovoltaic Program (DOE, 1991), of module efficiencies of 20% and system lifetimes of 20 years, would be met. This would still leave the matter of cost competitiveness in question in the mid-term, since the goals call only for a cost of PV electricity in the range $0.12–0.20/kW-hr by the year 2000.

By the mid-1990s, the best unit capacity costs of PV cells (alone) were over $3/W_p (peak watts) (DOE, 1991; Johansson *et al.*, 1993), with no indications of significant reductions in the near future. This reflects

PV-system levelized electricity costs exceeding $0.20/kW-hr, which is nearly twice the national price of conventionally generated electricity in the USA. It should be recognized, furthermore, that these costs are based on *peak* outputs, which are not equivalent to the capacity ratings (W_e, watts electric power output capacity) of conventional generation, because of the intermittency of the solar source. Peak capacity costs for solar generation must be lower than capacity costs of conventional generation to be competitive, because their lower capacity factors reflect a higher levelized cost (cents per kilowatt-hour) per watt of peak capacity. In addition, each watt of installed solar peak capacity cannot be included as part of the system **firm capacity** to meet all contingencies of demand because it cannot be counted on to supply the energy at any time on demand. Or, to put it all another way, the only way to make the intermittent solar source available more or all of the time is to provide energy storage and this increases the cost of the energy delivered (see Chapter 5). Consequently, the challenge remains for marketability of bulk PV power to be proven in any thing short of the long term.

Nonetheless, further progress can reasonably be expected, given the history of semiconductor research. With the PV industry, however, advances have seldom been achieved in the past on the time tables projected by many of the industry promoters (Merrigan, 1975; Maycock & Stirewalt, 1985; Hubbard, 1989). Such forecasts seem to be endemic in this and other evolving technology areas (Ascher, 1979; Katzman, 1984), and show no sign of abating (Thorton *et al.*, 1992; EPRI, 1991, 1994; CEC, 1994; *New York Times*, 1994, 1997; Maycock, 1995). Repeatedly, one has seen projections of levelized costs of PV electricity going from the present level of two to three times the level of conventional generation down to the competitive range within a decade or so. Projections of new thin-film production facilities have slipped nearly a decade (Kazmersky, 1997). Still, the goals of the US DOE Photovoltaic Program, for their long-term time frame, have some credibility in forecasting the cost of PV electricity falling to the competitive level of $0.05–0.06/kWhr over a time span ending in the year 2030.

It has long been the contention of PV advocates that mass production would bring dramatic reductions in unit costs (Stabaugh & Yergin, 1979; Maycock, 1995). Detailed studies have been carried out for large-scale production of various of the thin-film PV cells (Benner *et al.*, 1992; Zweibel & Barnett, 1993). Estimates have been made (Johannson *et al.*, 1993) that expansion of the annual production capacity of individual PV manufacturing plants could enable these plants to achieve significant production economies. Some analysts have suggested that PV cost estimates, for the purposes of investment decisions, be based on "vintage-levelized" calculations, which would have the benefit of the experience curves for mass production (Awerbuch, 1995). It would take annual production in individual plants at the peak megawatt output level to achieve economies of scale contemplated. For instance, it has been claimed that an increase from 1 MW_p to 10 MW_p annual production for a single plant would result in cost reductions of one third or more. Worldwide aggregate PV production was approximately 60 MW_p in 1993, with no single plant getting into the required range of production volume at that time (Stolte, 1994). Production was over 90 MW_p in 1996 worldwide (Curry, 1997), but facilities were "immature and unproved" (Kazmersky, 1997) for two of the promising cell types (polycrystalline Si and CdTe). Prospects for reducing production costs seemed poor for the mainstay (crystalline) silicon modules and production rates for a-Si are far below what is needed for competitive costs. The evident conclusion is that the basic technology is not yet mature for such expansion and that major investment in mass-production facilities would be risky.

It has often been proposed that government purchases could help to stimulate the industry into the major investments necessary to attain the economies of scale required, when the industry is deemed ready to achieve it. Being a "chicken and the egg" type of proposition, there has been hesitancy to undertake such ventures in the private sector and no coherent government policy has, as yet, been enacted to subsidize it. Some interest exists among the electric utilities for larger-scale PV installations, in supply support for existing grids and lines, but only if capital costs can be brought down to around $3/$W_p$ or below (EPRI, 1994). At a cost level of $3/$W_p$, a market potential for utility grid and line support has been estimated by the US utility industry at close to 9000 MW_p (UPG, 1994). If the unit costs of prototypes come down to approach this competitive range, it will be more likely that such production investments will be forthcoming. At the end of the 1990s, one larger-scale commitment has been made in the private sector (a 100 MW_p prototype system, manufactured and operated by one company), but this is speculative and dependent on government purchases or market guarantees (*New York Times*, 1994, 1997).

The prospects for large-scale generation by PV direct conversion, therefore, remain uncertain in the medium term. Whereas long-term prospects appear plausible because of the possibilities of innovations in semiconductors, forecasts of technical performance and costs cannot be accepted with much certainty regarding the timescale of their achievement. Nonetheless, research support for the National Photovoltaic Program, or equivalents, certainly appears warranted. It will only be through gains in our understanding of the fundamental physical properties of solar conversion that significant steps can be made toward the ultimate goal of direct solar conversion displacing conventional bulk generation of electricity to a significant degree. (See Chapter 11 for a general discussion of R&D policy.)

Meanwhile, demand for PV cells has grown in limited markets. Healthy markets exist for photovoltaics today in consumer electronics and for remote-site electric power. Worldwide, the annual production of PV cells (for all uses) grew 13% in 1993 (Brown *et al.*, 1997), with over 120 MW PV capacity manufactured per year by 1997, but this was not a major increase over the previous decade's growth rate. The consumer market (for pocket calculators, watches, etc.) is large ($400 M annually, worldwide) and dominated by Japanese firms manufacturing thin-film cells (Wilson, 1990). The low efficiencies of these products is unimportant as long as they give the small outputs (typically a few milliwatts) required. Their capacity costs typically would be $50/W_p$ or more, even with the benefits of mass production, but the costs of the tiny PV cells in these products is a small fraction of the overall manufacturing cost.

A major market for PV electric power, in the absence of large-scale grid connections, is for remote-site applications. These include a variety of uses (Strong, 1987; EPRI, 1991; Killebrew, 1991; Soderberg, 1992; Stokes & Bigger, 1992; Thorton *et al.*, 1992), for example remote water pumping, remote village (Third World) electrification, remote vacation cabin, utility-independent homes, utility-serviced stand-alone installations (where power lines are prohibitively expensive), and rugged-service (military, navigational, and communication-link) applications. In 1995, there was already an estimated aggregate of 25 MW_p in stand-alone PV capacity in the USA. There is a substantial listing of manufacturing firms and suppliers of complete systems supplying the PV cells and "balance-of-systems" equipment for these installations (PV Products, 1996). With unit costs running as low as $7/W_p$ for the cells and $10/W_p$ for entire systems, this makes economic sense for remote-site applications where long power lines are the alternative. For other remote or "rugged" applications, PV simply provides a technical alternative not otherwise available for reliable, low-maintenance service. At capacity costs as low as $3/W_p$, the market potential in the USA for remote PV applications has been estimated at over 200 MW_p (UPG, 1994); comparable amounts are estimated elsewhere (UN, 1992).

There is also growing interest in "building-integrated PVs" in architecture and "distributed generation" as utility connections (NESEA, 1996). An internationally connected group of architects has evolved a movement to design dwellings and commercial buildings that are both aesthetically pleasing *and* environmentally responsible. They are carrying out their designs not only by incorporating improved insulation, passive-solar features, and high performance lighting for the sake of efficiency but also are integrating PV panels into the construction (Strong, 1998). Thus, roofs and exterior walls can become solar collectors, with the durable sheets of PV semiconductors substituting for construction materials. These efforts have received the support of the IEA for a 5-year program and from various European governments (Schoen, 1998). In Germany, the program is called "The 1000 Roofs", giving 70% cost sharing for integrated PV installations. All of these programs are loosely part of a strategy, by enthusiasts, for diffusion of this technology into the world economy.

These programs are allowing the industry to gain experience in building and maintaining these systems during this era when PVs are not yet cost competitive on their own. A survey indicated that the prime motivation for a home owner or business to invest in these installations is that the buyer "wants PV" because solar "promotes sustainability". Investment costs in the building-integrated systems can be offset (partially) by arrangements with electric utilities of the type already available to non-utility generators in the USA and elsewhere whereby utilities buy the PV electricity at their own "avoided costs". The result for the customer is a "net purchase and sale" of electricity from the utility. The transaction can be done in several different ways, but the customer usually owes over a month's period because he purchases more than his PV system generates and because most utilities are allowed to charge the customer for the purchases at the retail rate. Retail prices, of course, are always higher than avoided cost rates; in California, for example, they are about three times as high. Some of this extra cost can be reduced by "green pricing" (see below).

In these "net metering" arrangements, the building-integrated PV has no battery storage and, therefore, uses the utility system as its back-up. Where there are extensive clusters of these systems, they can be of technical benefit to the utility's electric distribution lines. This synergism comes about when the PV generation occurs during the peak-load period of the utility, as it would in a sun-belt region for air conditioning, thus lightening the transmission loading on the lines. The Sacramento Municipal Utility District (SMUD) has such distributed generation, comprising hundreds of PV systems ranging in capacity from 4 kW to 30 kW each, totaling 5 MW (IRRC, 1995; NESEA, 1996).

Finally, demonstration projects have been operated successfully for central supply of solar–PV electricity to utilities since the late 1980s (Hubbard, 1989; Chowdhury, 1991; Thorton, *et al.*, 1992). The largest of these was a 6.4 MW_p PV system using single-crystal silicon cells. It was operated (1985–94) for the Pacific Gas & Electric Co. in Carissa, California, with low operational costs and low failure rate of the cells. SMUD also has a central PV generating plant, started in 1984 and now operating with a capacity of 2 MW_p.

In sunbelt regions, these solar systems are well matched to system load demands, not only for supplying power but also for reducing transmission and distribution requirements for peak-load periods. All told, there is at the end of the 1990s an aggregate of over 20 MW of grid-connected PV capacity in the USA, with about half of this in central-station installations such as at SMUD. The relatively small importance of this aggregate becomes apparent, however, when we recall that the total installed conventional generation in the USA is nearly 800 GW_e (800 000 MW_e). Further integration of PVs into utility systems in the USA is being promoted by industry groups like the Utility Photovoltaic Group (UPVG) and PVUSA (Photovoltaics for Utility Scale Applications). In Europe, only a few countries have significant fractions of their installed PV capacity connected to the grid at the present time, although it is an objective of the IEA sponsorship to strengthen those ties (Morse, 1997).

There are incentives for utilities to invest in PVs and other renewables. These come in the form of "green pricing" policies by state public utility commissions (Swezey, 1997) and municipal utilities (Asmas, 1998) in the USA. Such policies are part of modest promotional efforts for renewables, such as PVs, in view of their current non-competitive costs. In the USA, for example, levelized PV system costs of energy are presently in the range $0.25–0.50/kW-hr compared with utility rates (tariffs) in the range $0.05–0.18/kW-hr. The integration of PV sources into utility systems is also being carried out on a pilot basis in Europe and Japan (Firor 1993). The system engineering for power-grid connections has been the subject of studies in electric power engineering since the early 1980s (Chowdhury, 1988; Rahman, *et al.*, 1990). The economic strategies for integrating intermittent sources, such as solar and wind, into utility systems has also received significant attention (Kelly & Weinberg, 1993; Weinberg, 1994). In general, however, the intermittents can only penetrate to about 15% of a utility's installed capacity before reliability starts to be compromised. This assumes, of course, that electric-energy storage technologies are not available at competitive costs (see Chapter 5).

If PVs do become cost competitive, they can then be expected to penetrate the utility market to about this reliability limit. In that situation, they have the potential to provide about 15% of utility grid generation capacities in sunbelt regions. In the USA, this could mean an aggregate of over 50 GW_p nationwide, making up only about 7% of the overall installed capacity in all regions of the country. In the EC (southern regions), there is a potential, similarly constrained, for over 15 GW_p PV capacity, but this is only about 5% of installed capacity. Worldwide, there should be at least 100 GW_p potential for this grid-integrated PV capacity, again amounting to 5% or more of the present worldwide installed capacity.

In conclusion, it would appear that, in the absence of price competitiveness for PV systems and cost-effective energy storage, direct conversion of sunlight by photovoltaics cannot be expected to displace conventional electricity generation in a significant manner within the medium term. Prospects for a significant penetration of PVs into bulk power markets are uncertain in the longer term with regard to timing and will probably not occur until well into the 21st century (Kazmersky, 1997). In the meantime specialized and remote applications are expected to grow, making significant technological changes wherever they are applied.

References

Ahmed, K. (1994). Renewable energy technologies a review of the status and costs of selected technologies. *Energy Series, Technical Paper No. 240.* Washington, DC: The World Bank.

Aitken, D. & Bony, P. (1993). Passive solar production housing and the utilities, *Solar Today*, March/April, 23–26.

Anderson, B. (ed.) (1990). *Solar Building Architecture.* Cambridge, MA: MIT Press.

Andrejko, D.A. (ed.) (1989). *Assessment of Solar Energy Technologies.* Boulder, CO: American Solar Energy Society.

Anon. (1996). *Science*, 21, 1744–1745.

Ascher, W. (1979). *Forecasting: An Appraisal for Policy Makers and Planners.* Baltimore, MD: Johns Hopkins University Press.

Asmus, P. (1998). Power to the people the promise of green municipal aggregation. *Solar Today*, May/June, 26–31.

Awerbuch, S. (1995). New economic perspectives for valuing solar technologies. In: *Advances in Solar Energy,* Vol. 10, Ch. 1, ed. Boer, K.W. Boulder, CO:

American Solar Energy Society.

Babomf, J.D. (ed.) (1992). *Passive Solar Buildings.* Cambridge, MA: MIT Press.

Balcomb, J.D. (1992). Passive renewable energy – a promising alternative. *Solar Today,* Nov./Dec., 22–25.

Becker, N.D. (1992). The demise of LUZ: a case study. *Solar Today*, Jan./Feb., 24–26.

Beckman, W.A., Klein, S.A. & Duffie, J.A. (1977). *Solar Heating Design by the f-Chart Method.* New York: Wiley.

Benner, J.P., Olson, J.M. & Couts, T.J. (1992). Recent advances in high-efficiency solar-cells. In: *Advances in Solar Energy – An Annual Review of Research and Development*, Vol. 7, Ch. 4, ed. Boer, K.W., pp. 125–166. Boulder, CO: American Solar Energy Society.

Boer, K.W. (ed.) (1990). *Advances in Solar Energy – an Annual Review of Research and Development*, Vol. 6. Boulder, CO: American Solar Energy Society and New York: Plenum Press. (a) Fan, J.C.C *et al.*, High-efficiency III–IV solar cells, Ch. 3, pp. 394–426; (b) Swanson, R.M & Sinton, R.A., High-efficiency silicon solar cells, Ch. 4, pp. 427–484, (c) Zweibel, K. & Mitchell, R., CuInSe2 and CdTe scale-up for manufacturing, Ch. 5, pp. 485–579.

Boer, K.W. (ed.) (1992). *Advances in Solar Energy – An Annual Review of Research and Development*, Vol. 7. Boulder, CO: American Solar Energy Society and New York: Plenum Press. (a) Thorton, J. *et al.*, Recent advances in photovoltaic applications for utilities, Ch. 2, pp. 19–72; (b) Benner, J.P. *et al.*, Recent advances in high-efficiency solar-cells, Ch. 4, pp. 125–166.

Bronicki, Y.L. (1981). A solar-pond power plant. *IEEE Spectrum*, February. New York: Institute of Electrical and Electronics Engineers.

Brower, N. (1992). *Cool Energy – Renewable Solutions to Environmental Problems.* Cambridge, MA: MIT Press.

Brown, K.C. (1983). Re-examining the prospects for solar industrial process heat. *Annual Review of Energy*, 8, 509–530.

Brown, L.M., Renner, M. & Flavin, C. (1997). *Vital Signs, 1997.* New York: World Watch Institute.

Carlisle, N. & Christiansen, C. (1993). Solar domestic hot water as a DSM Measure. *Solar Today*, March/April, 19–20.

Carlson, D.E. (1989). Low cost power from thin film photovoltaics. In: *Electricity: Efficient End-use and New Generation Technologies and their Planning Implications*, eds. Johansson, T.B., Kelly, H, Reddy, A.K.N. & Williams, R.H. Lund, Sweden: Lund University Press and (1988) *Proceedings of the 8th International European Community Photovoltaic Solar Energy Conference.* Norwell MA: Kluwer Press.

Cassedy, E.S. & Grossman, P.Z. (1990, 1998). Solar-thermal technology. In: *Introduction to Energy Resources, Technology and Society*, 1st & 2nd edns. Cambridge, UK: Cambridge University Press.

CEC (Commission of European Communities) (1994). *The European Renewable Energy Study*, ECSS-EEC-EAEC. Brussels: European Community.

Chowdhury, B.H. (1988). Is central station photovoltaic power dispatchable? *IEEE Power Engineering Review*, December, 30.

Chowdhury, B.M. (1991). Optimizing the integration of photovoltaic systems with electric utilities. In: *Proceedings of the IEEE Power Engineering Society*, Summer Meeting, paper 91 SM329-3EC.

Cook, G. (1991). Photovoltaics. In: *The Energy Sourcebook A Guide to Technology, Resources and Policy*, eds. Howe, R. & Fairberg, A. New York: American Institute of Physics. (Assumes some knowledge of solid state physics in review.)

Cook, J. (ed.) (1989). *Passive Cooling.* Cambridge, MA: MIT Press.

de Laquil III, P., Kearney, D., Geyer, M. & Diver, D. (1993). Solar-thermal electric technology. In: *Renewable Energy – Sources for Fuels and Electricity*, Ch. 5, eds. Johansson, T.B., Kelly, H., Reddy, A.K.N. & Williams, R.H. Washington, DC: Island Press.

Curry, R. (ed.) (1997). *Photovoltaic Insider's Report.* Dallas, TX: Photovoltaic Insiders.

DOE (US Department of Energy) (1991). *Photovoltaics Program Plan, FY 1991-FY*, DOE/CH10093-92. Prepared by the National Renewable Energy Laboratory, Golden, CO.

DOE (1992). *Solar Industrial Program, Program Plan FY 1993-FY 1997.* Washington, DC: Office of Industrial Technologies, Office of Conservation & Renewable Energy (Draft, December 9 1992); Anon. (1992) solar process heat technologies begin to make their mark. *Industrial Energy Technology*, Nov.

DOE (1993). *Solar thermal electric, five year program plan, FY 1993 through FY 1997.* Washington, DC: Solar Thermal and Biomass Power Division, Office of Solar Energy Conversion (G.D. Burch, Director).

DOE (1995). *Photovoltaics: Program Overview Fiscal Year 1994.* DOE/CH10093-190. Golden, CO: National Renewable Energy Laboratory.

Duffie, J.A. & Beckman, W.A. (1974). *Solar Energy Thermal Processes.* New York: Wiley.

Dunn, P.D. (1986). *Renewable Energies: Sources, Conversion and Applications.* London: Peregrinus Press.

EPRI (Electric Power Research Institute) (1991). (a) *Amorphous Silicon Thin Films – The Next Step in Solar Cells, Report GS.3009.9.91*; (b) *Cost-Effective Photovoltaic Applications for Utilities and their Customers, Report GS.3005.8.91.* Palo Alto, CA: Electric Power Research Institute.

EPRI (1994). Palo Alto, CA. Emerging markets for photovoltaics. *EPRI Journal*, Oct./Nov., 7–15; Utilities planning more PV projects. *EPRI Journal*, Oct./Nov., 10.

Firor, K., Vigotti, R. & Iannucci, J.J. (1993). Utility field experience with photovoltaic systems. In: *Renewable Energy - Sources for Fuels and Electricity*, Ch. 11, eds. Johansson, T.B., Kelly, H., Reddy, A.K.N. & Williams, R.H., pp. 483–512. Washington, DC: Island Press.

Frankel, E. (1986). Technology, politics and ideology: the vicissitudes of federal solar energy policy, 1974–1983.In: *Energy Policy Studies*, Vol. 3, eds. Byrne, J. & Rich, D. New Brunswick, NJ: Transaction Books for the University of Delaware.

FSEC (Florida Solar Energy Center) (1979*). Proceedings of the Solar Energy Consumer Protection Workshop*, March, CONF-7805162, Cape Canaveral, FL.

Golob, R. & Bus, E. (1993). *The Almanac of Renewable Energy.* New York: Henry Holt.

Green, M.A. (1982). *Solar Cells, Operating Principles, Technology and System Applications.* New York: Prentice Hall. (A more detailed. treatment of solid state physics and semiconductor technology.)

Gupta, B.P. (1987). Status and progress in solar thermal research and technology. *Energy,* 12, 187–196.

Haggard, K. & McMillan, G. (1993). Straw bale passive solar construction. *Solar Today,* May/June, 17–20.

Hof, G. (1993). *Active Solar Systems.* Cambridge, MA: MIT Press.

Holl, R.H. & de Meo, E.A. (1990). The status of solar thermal electric technology. In: *Advances in Solar Energy,* Vol. 6, ed. K.W. Boer. New York: Plenum Press.

Howes, R. & Fainberg, A. (1991). *The Energy Sourcebook – A Guide to Technology, Resources and Policy.* New York: American Institute of Physics.

Hubbard, H.M. (1989). Photovoltaics today and tomorrow. *Science,* 244, 297–304.

Johansson, T.B., Kelly, H., Reddy, A.K.N. & Williams, R.H. (eds.) (1993). *Renewable Energy – Sources for Fuels & Electricity.* Washington, DC: Island Press. (a) Kelly, H., Introduction to photovoltaic technology, Ch. 6, pp. 297–336; (b) Green, M.A., Crystalline – and polycrystalline – silicon solar cells, Ch. 7, pp. 337–360; (c) Boes, E.C. & Luque, A., Photovoltaic concentrator technology, Ch. 8, pp. 366–401; (d) Carlson, D.E. & Wagner, E., Amorphous silicon photovoltaic systems, Ch. 9, pp. 403–435; (e) Zweibel, K. & Bornett, A.M., Polycrystalline thin-film photovoltaics, Ch. 10, pp. 437–481; (f) Firor, K. *et al.,* utility field experience with photovoltaic systems, Ch. 11, pp. 483–512; (g) Kelly, H. & Weinberg, C.J., Utility strategies for using renewables, Ch. 23, pp. 1011–1069.

Katzman, M. (1984). *Solar and Wind Energy – An Economic Evaluation.* Totowa, NJ: Rowman & Allenheld.

Kazmersky, L.L. (1997). Photovoltaics: a review of cell and module technologies. *Renewable and Sustainable Energy Reviews,* 1, 71–170.

Kelly, H. & Weinberg, C.J. (1993). Utility strategies for using renewables. In: *Renewable Energy – Sources for Fuels & Electricity,* Ch. 23, eds. Johansson, T.B., Kelly, H., Reddy, A.K.N. & Williams, R.H., pp. 1011–1069. Washington, DC: Island Press.

Killebrew, D. (1991). A Stand Alone Facility: The Cholla Recreational Site, *Solar Today,* Nov./Dec., 20–21.

Larson, R. & West, R.A. (1996). *Implementation of Solar Thermal Technology.* Cambridge, MA: MIT Press.

Levine, J.D., Hotchkiss, G.B. & Hammerbacher, M.D. (1992). Basic properties of the spheral solar cell. In: *Proceedings of the 22nd IEEE Photovoltaic Specialists Conference, 1991.* New York: Institute of Electrical & Electronic Engineers.

Lodhi, M.A.K. (1989). Collection and storage of solar energy. *International Journal of Hydrogen Energy,* 14, 379–411.

Lodhi, M.A.K. *et al.,* (1991). A solar bowl electric power plant for normal solar flux. In: *Energy and Environmental Progress,* Vol. B, ed. Verziruglo, T.N. Nova Science.

Lotken, M. & Kearney, D. (1991). Solar thermal electric performance and prospects – the view from LUZ. *Solar Today,* May/June, 10–13.

Marsh, W.D. (1980). *Economics of Electric Utility Systems.* Oxford: Oxford University Press.

Maycock, P. (1995). Photovoltaic technology, performance, markets, cost and forecast: 1975–2010. In: *Advances in Solar Energy, 10,* Ch. 8, ed. Boer, K.W., pp. 415–454. Boulder, CO: American Solar Energy Society; Maycock, P.

(1995). Photovoltaic technologies – a history and forecast 1975–2010. *Solar Today*, Nov./Dec., 26–29.

Maycock, P.D. & Stirewalt, E.N. (1985). *A Guide to the Photovoltaic Revolution.* Emmous, PA: Rodale Press. (Readable by a layperson.)

McConnell, R.D., Surek, T. & Witt, C.E. (1998). Progress in photovoltaic manufacturing. *Renewable Energy*, 15, 502–505.

Merrigan, J.A. (1975). *Sunlight to Electricity: Prospects for Solar Electricity by Photovoltaics.* Cambridge, MA: MIT Press.

Miller, B. (1998). Ozark Mountain Solar Home, *Solar Today*, March/April, 26–29.

Mitchell, R.L. & Surek, T. (1988). *SERI's Photovoltaic Research Program: A Foundation for Tomorrow's Utility-Scale Electricity*, Report No. SERI/TP 211–3354 for the US Department of Energy. Golden, CO: Solar Energy Research Institute.

Morse, F.H. (1997). Photovoltaic power systems in selected IEA member countries. In: Advances in *Solar Energy*, Ch. 5, ed. Boer, K.W. pp. 201–290. Boulder CO: American Solar Energy Society.

NESEA (Northeast Sustainable Energy Association) (1996). Building energy. *Proceedings of the 1st International Solar Electric Buildings Conference, Renew '96*, and *12th Annual Quality Building Conference*, Boston, MA. 4–6 March.

New York Times (1994). Solar power, for earthly prices – Entron plans to make the sun affordable, (A. R. Myerson, Business Day), 15 Nov. D1–2.

New York Times (1997). Solar power in big leap in California (National News), 19 May, B7.

Ohta, T. (1979). *Solar–Hydrogen Energy Systems.* New York: Pergamon Press.

PV Products (1996). Firms with catalogs or brochures advertising complete PV systems: (a) Photocomm, Inc., Grass Valley, CA; (b) ECS, Inc., Gainesville, FL; (c) Sunnyside Solar, Brattleboro, VT; (d) Uni-Solar, Troy, MI; (e) Atlantic Solar Products, Inc., Baltimore, MD; (f) Sunelco, Hamilton, MN; (g) Integrated Power Corp. (a Westington Company), Rockville, MD.

Rahman, S., Khallot, J. & Chowhury, B.H. (1990). Economic impact of integrating photovoltaics with conventional electric utility systems. *IEEE Transactions on Energy Conversion*, 5, 738–746.

Schoen, T. (1998). Building-integrated. PV in Europe. *Solar Today*, Jan./Feb., 22–24.

Shea, C.P. (1988). *Renewable Energy: Today's Contribution, Tomorrow's Promise, WW Paper 81.* Washington, DC: World Watch Institute.

Soderberg, P. (1992). Living beyond the top of the world. *Solar Today*, Sept./Oct., 11–13.

Solar Energy (1991). Special issue on solar ponds. *Solar Energy*, 46, 323–399.

SRRC (Solar Rating Certification Corp.) (1993–4). Washington, DC (an industry Group): (a) *Operating Guidelines for Certification of Solar Collectors, SRCC-OG-100, 1994* (update); (b) *Operating Guidelines and Minimum Standards for Solar Hot Water Systems, SRCC-OG-300, 1993* (under revision).

Stobaugh, R. & Yergin, D. (1979). *Energy Future.* New York: Random House.

Stokes, K. & Bigger, J. (1992). Reliability cost and performance of PV powered water systems. *Proceedings of the IEEE Power Engineering Society*, Summer Meeting.

Stolte, W.J. (1994). Cost of energy from utility-scale PV systems. In: *Advances in Solar Energy*, Vol. 9, Ch. 4, ed. Boer, K.W. Boulder, CO: American Solar Energy Society.

Strong, S.J. (1987). *The Solar Electric House*. Emmous, PA: Rodale Press.

Strong, S.J. (1998). A window to the future: building integrated photovoltaics. *Solar Today*, Jan./Feb., 35–38.

Swezey, B.G. (1997). Utility green pricing programs: market evolution or devolution? *Solar Today*, Jan./Feb., 21–23.

Thayer, B.M. (1994). Passive solar in the gloom belt. *Solar Today*, Nov./Dec., 23–26. (Architect Dennis Andrejko's home in Buffalo, NY.)

Thorton, J., DeBlasio, R. & Taylor, R. (1992) Recent advances in photovoltaic applications for utilities. In: *Advances in Solar Energy*, Vol. 7, Ch. 2, ed. Boer, K.W., pp. 19–72. Boulder, CO: American Solar Energy Society.

Trieb, F. (1997). Solar electricity generation: a comparative view of technologies, costs and environmental impact. *Solar Energy*, 59, 89–99.

Ullal, H.S. & Zweifel, K. (1991) Progress in the film solar PV technology. In: *Energy and Environmental Progress*, Vol. B, ed. Veziroglu, T.N. New York: Elsevier Science.

UN (United Nations) (1992). Prospects for photovoltaics – commercialization, mass production and application for development. *Advanced Technology Assessment System*, Issue No. 8, New York: United Nations.

UPG (Utility Photovoltaic Group) (1994). Photovoltaics: on the verge of commercialization. *Summary Report of UPVG Phase I Efforts*, June 1994. Washington DC: Utility Photovoltaic Group.

van Overstraeten, R. & Caratti, C. (eds). (1988). Photovoltaic power generation. *Proceedings of the Second Contractors Meeting*, Hamburg, September, 1987. Dordecht, the Netherlands: Kluwer.

Vant-Hull, L.I. (1992). Solar–thermal central receivers. *Solar Today*, Nov./Dec., 13–16.

Viceps, K.D. (1993). Southwest vernacular passive solar. *Solar Today*, May/June, 11–13.

Vories, R. & Strong, H. (1988). *Solar Market Studies: Review and Comment*. Golden, CO: Solar Research Institute.

Weinberg, C.J. (1994). The electric utility; restructuring and solar technologies. In: *Advances in Solar Energy*, Vol. 9, Ch. 5, ed. Boer, K.W. Boulder, CO: American Solar Energy Society.

Westby, R.D. (1994). The Federal solar thermal market: a second look. *Solar Today*, March/April, 19–21.

Wilson, A. (1990). Consumer products lighting the way. *Independent Energy*, 20, 64–66.

Williams, S. & Bateman, B.G. (1995). *Power Plays Profiles of America's Independent Renewable Electricity Developers*. Washington, DC: Investor Responsibility Research Center.

Winter, C.-J. Sizman, R.L. & Van-Hull, L.I. (eds.) (1991). *Solar Power Plants*. Berlin: Springer-Verlag. Also see *Proceedings of the 9–11 ISES Solar World Congresses*, 1989–91. London: Pergamon Press.

Wright, D.A. (1992). Durable free energy dwellings. *Solar Today*, Sept./Oct., 14–17.

Zangrando, F. & Johnson, D.H. (1985). *Review of SERI Solar Pond Work, SERI/TR-252-2322*. Golden, CO: Solar Energy Research Institute.

Zweibel, Z. & Bornett, A.M. (1993). Polycrstalline thin-film photovoltaics. In: *Renewable Energy – Sources for Fuels & Electricity*, Ch. 10, eds. Johansson,

T.B., Kelly, H., Reddy, A.K.N. & Williams, R.H., pp. . 437481. Washington, DC: Island Press.

Further reading

Bony, P. & Taylor, R. (1992). Solar opportunity in Northern Nevada. *Solar Today*, Nov./Dec., 20–21.

Brower, M. (1992). *Cool Energy – Renewable Solutions to Environmental Problems*, revised edn. Cambridge, MA: MIT Press.

Brown, L.R., Kane, H. & Ayers, E. (1993). Vital signs 1993 the trends that are shaping our future. In: *Energy Trends*, eds. Brown, L.R., Kane, H. & Ayers, E., pp. 52–53, New York: Norton for Worldwatch Institute.

DOE (US Department of Energy) (1987). *National Photovoltaics Program: Five Year Research Program, 1987–91, DOE/CH10093-7*. Washington, DC: Solar Thermal and Biomass Power Division, Office of Solar Energy Conversion.

EIA (Energy Information Administration) (1995). *Annual Energy Outlook 1995 with Projections to 2010*. Washington, DC: Energy Information Administration.

Gillett, W.B. (1985). *Solar Collectors: Test Methods and Design Guidelines*. Hingham, MA: Reidel.

Hill, J.E. (1979). *Experimental Verification of a Standard Test Procedure for Solar Collectors*. Washington, DC: National Bureau of Standards, US Department of Commerce.

Kearney, D. (1989). Solar electric generating stations (SEGS). *IEEE Power Engineering Review*, Aug., 4–8.

Klett, D.E. (1987). Solar water heating in the US. In: *Progress in Solar Engineering*, ed. Goswani, D.Y. Berlin: Springer-Verlag.

NREL (National Renewable Energy Laboratory) (1991–2). Program Overviews for the US Department of Energy, Office of Building Technologies. Golden, CO: Energy Information Administration. (a) *Passive Solar Design Strategies, NREL/SP-220-4330*, September, 1991; (b) *Passive Solar and Energy Technology Integration, DOE/CH10093-146*, June, 1992; (c) *Solar Heating, DOE/CH10093-124*, April, 1992.

Rahman, S., Khallot J. & Chowhury, B.H. (1988). A discussion on the diversity in the applications of photovoltaic systems. *IEEE Transactions on Energy Conversion*, 3, 738–746.

Rogers, E.M. & Shoemaker, F.F. (1971). *Communication of Innovation: A Cross-Cultural Approach*, 2nd edn. New York: Free Press.

Rogers, E. M. *et al.* (1979). *Solar Diffusion in California: A Pilot Study*. Sacremento, CA: California Energy Commission.

SERI (Solar Energy Research Institute) (1991). *Shedding a New Light on Hazardous Waste – Solar Detoxification, SERI/TP-220-3771, DE 91002135*, February. Golden CO: Solar Energy Research Institute.

Stone, J.L. (1993). Photovoltaics: unlimited electrical energy from the sun. *Physics Today*, Sept., 22–29.

Zweibel, K. (1990). *Harnessing Solar Power: The Photovoltaic Challenge*. New York: Plenum Press.

2 Biomass energy

Biomass energy, in the context of the present day industrialized world, means the use of natural organic resources to manufacture fuels. These resources include wood, crops, and organic wastes, which if used on a mass scale to displace the use of fossil fuels could lead to significant reductions in environmental pollution and net reductions in greenhouse gas accumulations. However, in order for biomass fuels to be utilized on a mass basis, production costs have to be lowered and major changes in agricultural and fuel production capabilities would have to be made. These achievements could only result out of a large-scale biomass R&D program at the national level.

The conception of biomass as an energy resource embodies a new view of the renewable natural resources that have been used by humans since they emerged on earth as a species. Wood as a fuel, of course, was historically the original biomass energy resource and is still a major source in the underdeveloped world. Peat, agricultural wastes, and animal dung, to a lesser extent, fall into the same category. The use of wood for fuels is thought to have contributed heavily to deforestation of the Levant in ancient times and of large parts of Europe in medieval times, as it has more recently in whole regions of the underdeveloped world. Yet, the modern ecological concept of forest resources views them ideally as renewable, not only for their replenishment but also more broadly for their benefit to our environment.

Wood and other biomass fuels fell into disuse in the developed world, along with hydro and wind power, with the advent of fossil fuels and industrialization. Today, even after some revived use of wood following the energy crisis, less than 3% of the primary energy output in the industrialized world comes from such fuels (Adler & Schwengels, 1991). There is, however, a newly growing realization of a large and diverse potential base of economic or near-economic biomass fuel sources, including managed forestries, energy crops, and a variety of available organic waste products.

The new conception of biomass fuel resources has arisen out of public perceptions of two major factors: the oil crises of the 1970s and the prospects for global climate change in the 1980s. First, the world has faced the prospects of insecure and ultimately exhaustible supplies of its principal fossil fuel – oil and, subsequently, it is beginning to face the prospects that its massive use of fossil fuels is likely to cause profound changes in global climates. This new conception of biomass fuels says several things: (i) alternative fuels can be developed that substitute for the (insecure and exhaustible) fossil fuels, (ii) these fuels can be made to substitute, in this way without contributing to climate change, and (iii) these fuels can be derived from renewable (inexhaustible) sources.

Modern-day biomass resources are seen to include wood, energy crops, and organic wastes, each appropriate in particular forms for mass use in either industrialized or developing societies. For instance, special species of trees and crops have been chosen for study and prototype projects because of their characteristics of growth, manageability, and environmental impact. Then, in order to convert these biomass resources into modern fuels, various biochemical processes are being considered, such as fermentation and anaerobic digestion, depending on the raw resource being utilized. In this chapter, we will review these various resources and processes as they are being investigated and utilized under limited government sponsorship in the US and other countries.

2.1 Biomass resources

Wood, used conventionally for direct combustion, accounts for about 7% of world energy production. Three quarters of this is consumed in developing nations by simply burning wood, harvested as needed for domestic uses. (A small percentage is converted to charcoal before burning (Earl, 1974).) In the developed nations, small fractions (3–4%) of annual energy production come from wood used in residential fire places and wood stoves (Brower, 1992). Further use of wood fuels has been recommended by the European Commission for use on the continent (IEA, 1994).

The use of wood and wood scraps as a captive fuel has long been a practice in the paper and wood-products industries of the USA and other industrial countries, accounting for a little over 3% of the primary energy consumption in the USA (Ranney,1986; NREL, 1993). Such use was expanded during the energy crisis of the 1970s (IGT, 1978), with part of the energy being used as process heat and part for electricity generation in co-generation schemes (see Fig. 2.1 for an example of such a plant operating in the State of Washington). Today, in the USA, there is an aggregate capacity of over 7 GW_e of such co-generation connected to the electric grids (Williams & Bateman, 1995); part of the generated output is sold to the utilities under the PURPA provisions at rates in the range $0.065–0.080/kW-hr (Bain & Overend, 1992; Brower, 1982). In addition, there is considerable remote-site use of wood in these same industries for processing and captive co-generation (i.e. not feeding utility grids). All together, the paper and wood-products industry uses well over half of the wood-fuel consumption in the USA.

In recent years, several hundred megawatts of electric output of wood-fired co-generation have been added annually in the USA, even after the demise of federal tax credits in 1989. The rate of further expansion will depend on price competitiveness of these plants. Wood-burning power plants of conventional designs have been estimated worldwide to be able to generate electricity at a cost of about $0.075/kW-hr (Ahmed, 1994); by comparison, a novel "whole-tree burner" design is projected to bring this

Figure 2.1
Wood-fueled
co-generation plant in
Washington State.
(Courtesy of the
National Renewable
Energy Laboratory.)

cost under $0.05/kW-hr (Brower *et al.*, 1993). Ultimately, the competitive-ness of wood-burning electricity generation will depend variously on achieving higher thermal efficiencies in the new plants, a rise in the prices of fossil fuels (e.g. coal), or the possible resumption of tax credits for renew-able-energy investments.

While wood-fired co-generation has some promise, it hardly represents a technology innovation with potential to influence energy markets much beyond the paper and wood industry. There have been innovative propo-sals (Bain & Overend, 1992) to use biomass feedstocks instead of coal in a promising new technology similar to co-generation: this is the "integrated gasifier/combined cycle" (IGCC) generation. This could have some im-pact, albeit limited, on the electric sector. In order for wood resources to have a wider impact, technological means must be found to create gaseous or liquid fuels from them for wider uses. However, if such processes are developed for mass use, then means of managing the wood resources in a renewable manner must be devised if sustainability is to be achieved (Kishor & Constantino, 1994; Evans, 1997). In addition, environmental issues associated with fuel-burning power plants, while they might not be as severe for wood fuels as for coal, would have to be addressed.

Ideally, wood or any of the energy crops that have been proposed can be managed for a carbon cycle that results in no net generation of carbon dioxide (Fig. 2.2) and thus moves towards sustainability for global-warm-ing concerns. Actual net emissions of carbon dioxide as low as 5% are thought to be feasible (Overend, 1998). For wood, the use of fast-growing species, such as poplar, sycamore, and silver maple, have been proposed (Ranney, 1986; NREL, 1992; DOE, 1993; Tuscan, 1998) for wood produc-tion in "short-rotation" tree farms. Laurel (*Cordia alliadora*) has been

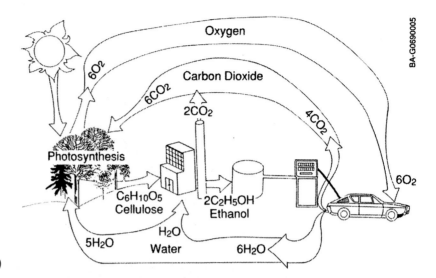

Figure 2.2
The carbon cycle for biomass fuels. (Courtesy of the US Department of Energy.)

proposed in (tropical) Latin America (Kishor & Constantino, 1994). In addition to high productivity of wood (as much as 10 tons annually per acre), new growth of trees has a rate of carbon fixation that is over three times the rate of mature growth. This carbon-fixing rate for new growth is 4–5 ton CO_2 per acre (9–11 mt/ha) annually, which roughly matches the CO_2 production of the combustion of this wood. (Woods are about 50% carbon composition by weight (Dunn, 1986).)

Short-rotation, intensive-culture (SRIC) tree plantations have been shown to be highly productive (Ranney, 1986; Strauss & Grado, 1991). SRIC techniques use fast-growing tree species planted at close spacings and harvested at short intervals to maximize the productivity per acre of land used. An example of a 4-year-old plantation of sycamores is shown in Fig. 2.3a. Spacings of 2.5–1.5 m (allowing 650–1800 trees per acre) and harvesting intervals of 4–7 years have produced up to 5 tons (dry) (4530 kg) wood per acre, as an annual average, in temperate climates. Either new trees can be planted after harvesting or, in some species, the stumps are allowed to re-sprout, in a process called coppicing (Figs. 2.3b,c). Coppicing can be done at even closer planting spacings and quick growth times can still be achieved. The various species appropriate for SRIC in the USA are shown in Fig. 2.3d, where their regions of use are indicated. SRIC is successful for increasing productivity in hybrid willow in short rotation cycles (Tuscan, 1998) and for reducing the need for pesticides (Sage, 1998). Larger yields are also possible in tropical climates, for example eucalyptus cultivation in Hawaii or southern Florida (IGT, 1978).

In Europe, which is densely populated, prospects for wood feedstocks for biofuels are limited by land use. Austria, Finland, and Sweden have the largest forest resources, which presently supply 12–15% of their own primary energy demands (Asplund, 1996). Wood supplies only about 2% of demand in the other EU countries , although France and Germany are

(a)

(b)

Figure 2.3
Short rotation tree farming. (a) Spacing arrangements in a sycamore plantation. (b) Coppicing in SRIC farming: regrown willow shoots. (c) Row of regrown coppiced silver maple. (d) Regional distribution of tree species. (e) RSIC tree harvesting in Canada. (Courtesy of the Oak Ridge National Laboratory.)

(c)

(d)

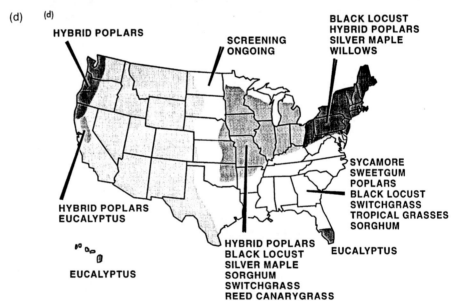

(e)

Figure 2.3 (cont.)

thought to have potential for greater supplies. There is no principle wood harvest for fuel wood at the present time in Europe, however; only logging residues and lumber scraps are used. Poplar and willow are thought to be good prospects for sustainable wood cropping in Northern Europe (van der Bijl, 1996; Overend, 1998), as they are in North America, and eucalyptus in the more southern climes of Portugal, as in the United States southwest and in South America.

Productivity, production costs, and sustainable production are competing objectives in wood or any of the biomass feedstock operations. As in any agriculture, of course, more fertilizers and irrigation can increase annual production, but at an added production cost and possibly to the ultimate detriment of sustainability. The approach being taken in energy-crop R&D is to find plant species, site locations, cultivation methods, and harvesting techniques that maximize production and minimize costs, within constraints of sustainability. The SRIC program is an excellent example of this approach.

Efficient harvesting techniques have been developed for short-rotation, woody-crop (SRWC) plantations (Ranney, 1986; Patterson, 1994). Using equipment similar to farm harvesting machinery (Fig. 2.3e), the trees are continuously cut, accumulated, and bundled as the tractor moves along a row of saplings. Since harvesting accounts for at least a third of the overall cost of plantation production (Strauss & Grado, 1991) this efficiency is critical. Such equipment can cut about 850 trees per hour.

The net outcome of these cultivation and harvesting techniques has to be high land productivity at the lowest sustainable cost. Land productivity is important in containing cost since land rent or amortization must always be assumed in any form of agriculture (Evans, 1997). Land productivity should be at least 10 tons per acre (22mt/ha) annually, with each ton of biomass feedstock costing $35 or less to produce: a level achievable in the USA and in Europe (IEA, 1994). A useful comparison can be made with such costs in that a unit energy cost of approximately $3.50/MBTU ($3.69/GJ) is found for such a biomass feedstock when its heating value is in the characteristic region of 10 MBTU/ton (11.6 GJ/mt). Such a unit energy cost is, of course, near competitive with fossil fuels such as oil and natural gas.

Energy crops make up another major resource for biomass fuels. Whereas crop residues have been used historically as direct-burning fuels and are still used today in the developing world, there is now the technological prospect for mass conversion of whole crops to liquid or gaseous fuels. Starchy crops, such as corn and sugar cane (Fig. 2.4a), are proven feed stocks for ethanol. Wheat and sugar beets are being evaluated in Europe (IEA, 1994; EUREC, 1996). Now, even herbaceous crops such as grasses (Figs. 2.4b,c) and legumes can be processed into alcohols. Recently reported research in the USA opens, for the first time, possibilities for conversion of cellulosic materials to ethanol (Magee & Kosaric, 1985; Vallander & Eriksson, 1990; Lynd, 1989; Lynd et al., 1991). This not only increases the alcohol yield from the starchy crops, by allowing conversion of the stems and leaves, but will also allow the inclusion of resources that are entirely cellulosic, such as wood, perennial grasses, and legumes (more on this in following sections). This basic discovery has stimulated a new R&D program on herbaceous energy crop production (HECP).

Another plant group that holds possibilities for energy crops is the genus *Euphorbia*, which contain (hydrocarbon) latex emulsions (Calvin, 1984). Nobel laureate Melvin Calvin has proposed investigation of this family of plants, which are related to rubber trees and include the milkweed. Some species of *Euphorbia* can be grown in semiarid climates on land otherwise unusable for agriculture (Calvin, 1979). The oils extracted from such plants are similar in chemical composition to petroleum and can be cracked and refined directly into fuels and lubricants; hence its advocates call it "biocrude". Many of these plants also contain fermentable sugars (for alcohol fuels) and burnable residues (for use as captive fuels in processing). Further investigation is required, however, including possible toxic by-products of processing.

It has been estimated by the US DOE (Bull, 1991; Lynd 1989; Lynd et al., 1991; DOE, 1993) that nearly 200 million acres (81 million hectares) of land in the USA (about 10% of the total) could readily be made available for energy crops, using idled food-crop lands plus some pasture range and forest lands that could support energy crops. It has been estimated that energy crops in the US Upper Midwest alone have the potential to supply 22% of the total primary-energy needs of those 12 states. The distribution of idled lands in the US is mostly over the eastern two thirds of the country, with the greatest concentrations in the Midwest and South (Fig. 2.5a). The crop lands presently used in the USA for food production total a little over 300 million acres (122 million hectares). Furthermore, the Oak Ridge National Laboratory (USDA, 1992) has estimated that 392 million acres

(a)

(b)

Figure 2.4
Energy crop farming.
(a) Energy cane field.
(b) Switchgrass
harvesting. (c)
Switchgrass harvested
bales. (Courtesy of the
Oak Ridge National
Laboratory.)

(c)

Figure 2.4 (cont.)

(157 million hectares), of the total crop lands of all categories, are capable of growing energy crops without irrigation (Fig. 2.5b). The US Department of Agriculture has even recommended that the 150 million acres (61 million hectares) of idled food-crop lands be re-employed for non-food farming (Lynd, 1989; Lynd, *et al.*, 1991). The prospective energy-crop lands have the potential to supply over 25 QUAD per year (26 EJ/year) of primary energy for the USA, nearly one third of the nation's present total demand. The Congressional Office of Technology Assessment (OTA) had earlier estimated 10 quads (10.6 EJ) from wood and 5 quads (5.3 EJ) from grasses alone could be obtained (Antal, 1983). Recently, the National Renewable Energy Laboratory made a similar estimate for cellulosic feedstocks (Saha & Woodward, 1997).

The biomass resource situation is considerably less promising in Western Europe, where all usable agricultural land is currently utilized for food production (Lehmann *et al.*, 1996). About 23% of the over 300 million hectares (740 million acres) of the EU-15 countries is in a degraded condition, some of it beyond feasible recovery, although there is a claim that 30–40 million hectares of this could be made available for future energy biomass production (Hall, 1994; EUREC, 1996). The acreage used for food agriculture in these 15 countries is close to 400 million acres (160 million hectares), comparable with that for the same use in the USA. With a population of about 340 million in the EU-15, this also gives a comparable food production per capita with the USA (256 million population, with 300 million acres (121 milliion hectare) cultivated for food). Only with an adjustment in the use of arable land away from food would significant acreage be available for energy crops, but this would likely require adjustments within the Common Agricultural Policy of the EU (Hall *et al.*, 1993). Even if all of this was achieved, it would not be expected to supply more than 10% of the primary energy demand of the EU-15 countries. The only likely resource areas for possible use as energy feedstocks are then

(a)

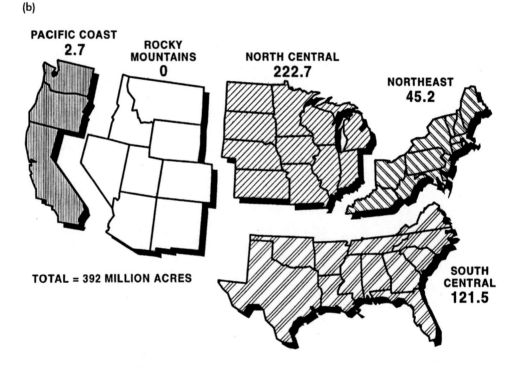

Figure 2.5
Energy cropland resources. (a) Long-term potential for USA energy crop production. (b) Land capable of producing energy crops without irrigation. (Courtesy of the US Department of Energy.)

(b)

unprotected forest areas (a little over 250 million acres (100 million hectares)), but use of any significant portion of this would run contrary to recommended sustainable use (Lehmann *et al.*, 1996).

Prospects for biomass energy resources in other industrial countries vary widely. Japan, of course, with a population density over ten times that of the USA and higher than most European countries, has to import over 12% of its food – a higher proportion than most Western European countries. Australia, by comparison, has vast land areas, which should have prospects similar to those of the American West, even with portions that are arid. Canada, likewise, has large land areas, but only the southern strip is suitable for crops, leaving wood as the prime feedstock for biofuels (Brown *et al.*, 1997).

In general, of the developed countries, the USA, Canada, Argentina, and Australia have the best prospects in arable land availability, ranging from 0.7 ha/capita (1.7 acre/capita) (USA) to 2.9 ha/capita (7.2 acre/capita) (Australia) (Pimental & Giampietro, 1994). Of these, the USA is in the best position, when one includes factors such as temperate climate and average rainfall. Europe and Japan, by contrast, have the least land availability, because of population density, with arable land running mostly less than 0.2 ha/capita (0.5 acre/capita) in Europe down to 0.03 ha/capita (0.08 acre/capita) in Japan. Correspondingly, the USA, Canada, Australia, and Argentina are presently the major food exporters (over 80% of world cereal exports), while Japan and Europe must rely on agriculture with heavy fertilizer and energy inputs for high land productivity, and imports in addition, to feed their people.

The underdeveloped countries are, in general, not in situations to consider agriculture for energy crops for the foreseeable future, having to struggle to raise enough food for their populations. Countries, such as Bangladesh, China, and Egypt, are themselves densely populated, with arable lands of 0.08 ha/capita (0.2 acre/capita) or less. Other LDCs, such as those in Africa, having significantly more available land and ample rainfall, suffer from low land productivity and the lack of energy and fertilizer inputs to increase it (Pimental & Giampietro, 1994). Currently, efforts are under way to sustain, and hopefully improve, the per capita food production in the face of population growth (WRI, 1996; Keeney, 1996; Anon., 1997b). The prospects for the utilization of large land areas in the LDCs for energy feedstocks, therefore, will most probably have to come after the development of biofuels in the industrial world. Until such a time, it is to be hoped that resources such as arable land and forests will not have been depleted in the hunger for development and the rising demand for food (Brown *et al.*, 1997). Even when industrial development does come, there is the threat that reduced arable land per capita will place some of these countries into food-import dependency, such as exists already in Japan, the Koreas and Taiwan (Gardner, 1996). Russia and other parts of Eastern Europe are likely to be in a similar position for the first part of the 21st century.

The inclusion of cellulosic energy crops would not only widen the scope of biomass resources worldwide but also would open possibilities for reducing the production requirements for fertilizer and energy inputs of conventional (food) crops such as corn (Lynd, 1989; Lynd, *et al.*, 1991; NREL, 1992). Perennial switchgrass, for example, requires much less fertilizer for its (rapid) growth than corn since soil nutrients are mostly retained after harvesting the grass. In addition, grass is harvested more efficiently (Fig. 2.4b,c) than crops such as corn. Also, wood crops require less than half of the energy input (fertilizer and fuel) in growing and harvesting than corn, measured as a fraction of the output biofuel energy content. The energy input requirements for ethanol from corn are about

40% for the feedstock itself and the high production costs ($80–180/ton or $88–198/mt) (IEA, 1994), both of which make it of questionable value as a feedstock for processing into alcohol fuels.

In addition to the technical requirements for biofuel conversion, closely related market economic factors must be considered when comparing biomass feedstocks (Lynd *et al.*, 1991). By-product credits play a major role in holding down the cost per gallon of corn-produced ethanol at the present levels of production. Much higher levels of production than those achieved in the mid-1990s will be required for ethanol to make a significant impact nationally as a transportation fuel, however. If such a level of production were attempted, using corn as the feedstock, the markets for corn by-products (USDA (US Department of Agriculture), 1989, 1993) would be saturated and the co-product income would also not be available. In addition, the price of corn and corn co-products fluctuates greatly as a result of the food commodity and livestock feed markets even under current market conditions (Wyman *et al.*, 1993). During a 4-year period in the 1980s the price of corn fluctuated nearly 3:1 and co-products 5:2 in the USA. In 1995, the market price for corn rose from $ 2.60 to $4 per bushel, forcing some ethanol processors to cease operations (DOE, 1996a).

Energy crops which are also food crops, such as corn and sugar, create the undesirable situation of coupling the food and fuel markets (Hall & de Groot, 1986). Prices in one market will influence prices in the other, and both would be dependent on annual farm production fluctuations. Such a cross-market dependency on either feed stocks or co-products would not be nearly as great for feedstocks such as wood or non-food crops and certainly not for municipal solid wastes (MSW). Earlier concerns over the impending competition between food and energy crop cultivation (Garg, 1987) would, therefore, appear to be alleviated by the use of non-food energy crops, at least in the USA where the additional crop lands are available. In Brazil, significant cross-impacts were experienced with the massive program for producing ethanol derived from sugar cane. In the EC, agriculture policy is in nearly constant dispute and the entrance of energy crops would undoubtedly intensify the existing tensions (Patterson, 1994). However, a major consideration in the EC program for biofuels is the maintaining of agricultural lands in production and sustaining rural employment (IEA, 1994). The European biomass programs do not seem to emphasize non-food crops, such as grasses, as feedstocks, putting emphasis more on the plant-oil crops such as rape seed (Korbitz, 1998).

In any of the energy crops, whether starchy, herbaceous, or woody, the soil nutrients must be maintained, just as they must be for any form of agriculture (USDA, 1992). If fertilizers are used, they become part of the cost of production, both in monetary terms and in consideration of the "net energy" (Henderson, 1988) of the fuel-conversion process. If the fertilizers are derived from fossil sources (e.g. natural gas) then the gains of sustainability become partially compromised. Water pollution (aquifer and tributary) can also become a hazard with fertilizers, as elsewhere in agriculture. Account must be taken of aquifer resources in order to avoid depleting ground waters. Land-conservation practices must be adhered to

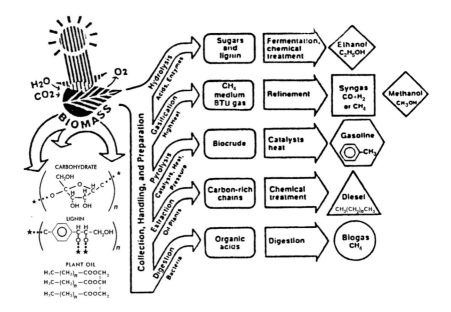

Figure 2.6
Biofuel processes and
products.(Courtesy of
the US Department of
Energy.)

for the crops and accepted silvaculture practices in tree rotations and
harvesting must be ensured to avoid erosion and soil degradation (Maser,
1994; van der Bijl, 1996). Small parts of the forest should be left after
harvesting for soil nutrients and moisture retention. (One criterion used is
to provide at least 50% ground cover.) Similar considerations apply to
other energy crops and are the subject of ongoing study (USDA, 1993). It
is generally recommended, for example, that a minimum of 1 ton/acre of
crop residues be left in the fields after harvest. Research is needed on these
ecological issues and others, including biodiversity (Boyle & Boyle, 1994),
before biofuels can be considered as truly sustainable replacements for
fossil fuels. More discussion on these biomass feedstock issues, as well as
the issues concerned with biofuels processing, is given at the end of this
chapter.

Biomass wastes are also a significant resource for fuels (Bull, 1991).
Wood wastes from the forest-products industries have already been mentioned in connection with direct combustion. However, wood wastes are
also usable for processing into fuels (Lynd, 1989; Lynd *et al.*, 1991; Schell
et al., 1992): in solid form (e.g. charcoal), as liquids (e.g. by pyrolysis), and
possibly even gaseous forms (such as methane and carbon monoxide), as
will be discussed later. Agricultural wastes, including corn stover and
wheat straw, are also possible large resources for such processes (Magee &
Kosaric, 1985). The potential primary energy for the US energy demand
from these two resources – wood and agricultural wastes – has been
estimated at 1.7 quads (1.8 EJ) and 1.4 quads (1.5 EJ), respectively,. In
both cases, however, the overzealous removal of waste matter on forest or
agricultural lands would be counterproductive for soil fertility ; the optimal practice evidently needs to be worked out.

By contrast with the other biowaste categories, MSW is waiting for full

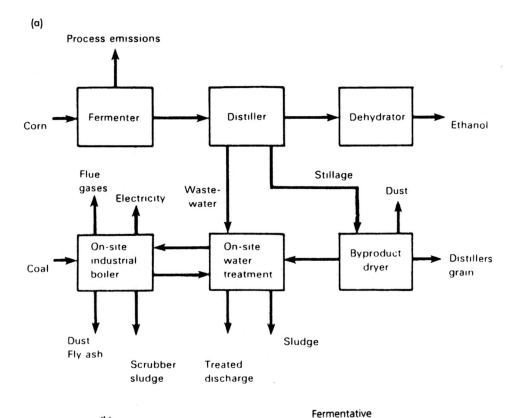

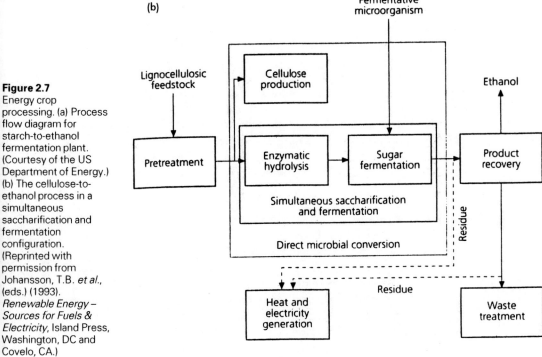

Figure 2.7
Energy crop processing. (a) Process flow diagram for starch-to-ethanol fermentation plant. (Courtesy of the US Department of Energy.) (b) The cellulose-to-ethanol process in a simultaneous saccharification and fermentation configuration. (Reprinted with permission from Johansson, T.B. *et al.*, (eds.) (1993). *Renewable Energy – Sources for Fuels & Electricity*, Island Press, Washington, DC and Covelo, CA.)

utilization, presenting solely problems of environmentally safe disposal. MSW compositions are typically about 40% paper plus another 40% of other combustibles, equivalent in heating value (HV) to that of wood or peat (about 10 MBTU/ton or 11.6 GJ/mt), which is about half the HV of coal (Cheremisinoff & Morresi, 1976). The promising processes for MSW-derived fuels are gasification (methane) and alcohols (e.g. ethanol), There is potential in the USA for production of not quite 1 QUAD primary energy from MSW (Williams & Bateman, 1995) of which a fraction could be in the form of these alternative gaseous and liquid fuels. MSW would not be a major contribution to the national energy resources in the USA, having an aggregate annual energy content, nationwide, less than one twentieth that of wood (Brower, 1992). Nonetheless, utilization of the masses of MSW organic materials nation-wide would be a major contribution to an environmental dilemma that faces virtually all cities and townships in the USA and other countries today.

The conversion of MSW organics to fuels would provide alternatives to "trash to energy" schemes of recent years, which have attempted direct combustion to raise steam electricity generation (Cheremisinoff & Morresi, 1976). These mass-burn projects have aroused widespread public opposition, because of reports of carcinogenic emissions (Penner *et al.*, 1987; Brower *et al.*, 1993). While many of these MSW problems in the USA are institutional, the general need for energy as a by-product of disposal arises out of financial pressures on municipalities to offset the costs of disposal. In other countries, waste-to-energy plants have attained more widespread adoption than in the USA (Brown, *et al.*, 1997; Warmer Campaign, 1991–94). Switzerland burns over three quarters of its household wastes, while Sweden burns over half. Sweden and other European countries stopped building further mass-burn plants in the late 1980s while they investigated the risks of dioxin emissions and set air-quality standards on all emissions.

The proposed MSW-to-fuels schemes, such as those involving anaerobic digestion, hydrolysis, or fermentation processes, would provide technological alternatives to mass combustion. The new possibilities of converting cellulosic materials to alcohols expands the fuels possibilities with MSW greatly. Finally, some limited resources of methane, as naturally generated by anaerobic decomposition, may be tapped from MSW landfills. These various possibilities will be discussed in the following sections.

2.2 Liquid biofuels

Biomass fuels hold a particular promise for replacement of liquid fossil fuels for transportation: namely, gasoline and diesel fuel. If biomass liquid fuels such as alcohols can be produced competitively, then the technological and economic advantages of the ICE can be carried over into the new era of sustainable energy. Currently, petroleum-based transportation fuels emit 27% of the anthropogenic CO_2 in the USA. Presently, 100 areas of the country are in non-compliance with ambient air-quality standards for ozone and 40 areas for carbon monoxide, mostly as a consequence of automotive emissions (Vallander & Eriksson, 1990; Lynd, 1989; Lynd, *et al.*, 1991).

Liquid biomass fuels, having energy contents comparable to petroleum, will retain the advantages of transportability and storability for supplies and stocks of the oil industry (Cassedy & Grossman, 1990, 1998). These were some of the advantages cited for the coal-based synfuels programs of the 1970s. Equally important are the technological performance features of the ICE, which have yet to be challenged by the electric vehicle in any of its forms. Weight per unit of energy stored or power delivered determines limits on the range and speed of the alternative vehicles, which presently do not compete with conventional automotive technology and have shown few prospects for doing so over recent decades. (See the discussion of electric vehicles in Chapter 5.)

Biomass may be converted into liquid or gaseous fuels by either bio-chemical or thermochemical processes (Fig. 2.6). The biochemical processes are fermentation for ethanol and anaerobic digestion for a methane-based syngas. The thermochemical processes include pyrolysis, possibly followed by catalytic and shift processes of the type used in coal-based synfuels (to be discussed in a later section).

Fermentation has been used historically, directly with sugar feed stocks and with starchy feed stocks following hydrolysis to make alcoholic beverages. Fermentation (again with sugars or starches) for ethanol fuels has been operational throughout the 20th century, going to mass production in Brazil since the 1980s (Garg, 1987; Goldemberg *et al.*, 1993). Brazil, fermenting sugar (sucrose) from sugar cane, has an annual ethanol production exceeding a half billion gallons (2 billion liters). Pilot production of ethanol has been carried out in the USA since the oil-crisis period, using corn as the feed stock (see a process schematic in Fig. 2.7a). During that period, the US government sponsored over 200 alcohol fuels facilities, most of them small scale. The largest plant was the Archer Daniel's Midland plant in Illinois, which reached an annual production somewhat over 50 million gallons (189 million liters) ethanol in the 1980s from a corn feed stock. The net benefits of the corn-to-ethanol program have been challenged, however, regarding environmental impacts, energy savings, and cost competitiveness (Schippler & Sperling, 1994) (more on these questions below).

The total projected output of the US government ethanol program was originally planned to rival the Brazilian output from sugar cane. Indeed, with the help of government incentives and technology gains, by 1992 ethanol annual production from corn in the USA reached 1.1 billion gallons (4.2 billion liters) and is on a commercial basis (USDA, 1992, 1993; DOE, 1994a). This, however, is sufficient to displace less than 1% of the nation's automotive consumption of petroleum. The (corn-based) industry has estimated that production costs can be brought down to the near competitive range of $1.08–1.95/gallon ($0.28–0.52/liter) in the near term, thus suggesting great expansion of production when this point is reached. However, if mass production of corn-derived ethanol were considered to make a significant impact on a nationwide scale, the cross-market effects of the corn feedstock with foods would have to be assessed, along with the other issues mentioned above.

More recently, means have been found to break down cellulosic matter into fermentable sugars by hydrolysis processes aided by acids or newly discovered enzymes (Magee & Kosaric, 1985; Lynd, 1989; Vallander & Eriksson, 1990; Lynd et al., 1991; Schell et al., 1992). This expands the possible feed stocks for ethanol to include wood and non-starchy crops. Fermentation for alcohol can take place on any feed stock that can be converted into a sugar, such as sucrose or glucose. Fermentation is then followed by distillation (Fig. 2.7a). In the case of corn, the starch is readily dissolved into a fermentable sugar (glucose) by hydrolysis, aided by long known enzymes. Cellulosic matter, however, does not easily break down into fermentable sugars. Until recently, prospects for converting cellulosic matter, such as woods, grasses, or the fibrous matter of agricultural products (e.g. corn stalks), seemed to be nil.

Hydrolysis of cellulosic materials into fermentable sugars is not as straightforward as the process from starches. Acid catalysts or special enzymes are required in the chemical processing, which must be preceded by treatments to make the lignocellulosic matter more digestible. These pretreatments include mechanical comminution (chipping, grinding, or milling) plus further breakdown by explosive decompression techniques or chemical pulping processes (Vallander & Eriksson, 1990; Schell et al., 1992). Acid catalysts are expensive for the concentrations needed to give high sugar yields, and, therefore, enzymatic hydrolysis appears to be the leading candidate for high-yield production and cost reduction.

A significant consideration in processing cellulosic matter for fermentation is achieving the hydrolysis of both components: cellulose and hemicellulose. Cellulose is a polymer of glucose that can be hydrolyzed into its basic form glucose (a fermentable sugar). Hemicellulose, however, is a polymer of several other sugars, mainly xylose, which is the main product of its hydrolysis. Xylose, however, presents more obstacles to fermentation and catalysts must be used. It is this factor in the breakdown of the hemicellulose fraction that becomes critical in determining the productivity and costs of alcohol from both cellulosic crops and wood. Herbaceous (cellulosic, but not woody) crops are about 45% cellulose and 30% hemicellulose, whereas hard woods are 50% and 23%, respectively. Consequently, the conversion of the hemicellulose fractions is significant for both types of energy crop. Similar figures hold for agricultural wastes.

It has been found that cellulosic feed stocks (e.g. wood chips) can be fed, following pre-treatment (e.g. pulverizing), into a process combining hydrolysis and fermentation (Vallander & Eriksson, 1990, Schell et al., 1992, Wyman et al., 1993). This combined process is called "simultaneous saccharification and fermentation" (SSF) (Fig. 2.7b). The cellulose is subjected to hydrolysis, aided by enzymes, in the presence of yeasts that ferment the glucose as fast as it is produced. Meanwhile, the xylose – also liberated by hydrolysis – is fermented using particular strains of yeast as catalysts with carefully controlled aeration. These newly developed techniques appear to be opening possibilities not evident previously for high-productivity conversion to ethanol from cellulosic matter.

The remaining organic matter in these cellulosic feed stocks is lignin, which is not readily converted by any of these hydrolysis processes. Lignin typically comprises about 22% hardwoods and 15% herbaceous crops. The lignin residues of these ethanol processes can, however, be used as a captive fuel to provide the heat required for distillation or for cogeneration, thereby improving the energy ratio of the production process (Wyman *et al.*, 1993).

Alcohols, such as ethanol, have both advantages and disadvantages as automotive fuels (Lynd, 1989; Lynd *et al.*, 1991; Schell *et al.*, 1992). They may be used either as pure alcohol or as blends with gasoline. Ethanol, for instance, has a higher octane than gasoline, allowing higher compression ratios in the engine. This allows operation with less engine knock and can lead to engine designs with high fuel economies. A high compression ratio can also give a greater delivery of power for acceleration. However, ethanol has a lower energy content per unit of volume, which means a reduced driving range for the same fuel tank capacity. With optimal engine design for ethanol, 80% of the range of comparable gasoline engines can be achieved. Finally, alcohol, being less volatile than gasoline, makes for poor starting characteristics in cold weather, unless aided by prewarming.

Alcohol fuels – ethanol in particular – are superior to petroleum fuels for the environment (Office of Technology Assessment, 1990; Archer, 1992; Brower, 1992). The lower combustion temperatures of alcohols result in less emission of the nitrogen oxides (NO_x). This feature, plus the absence of aromatic fumes from gasoline evaporation, leads to reduction of photochemical smog and ozone production at ground levels. Evaporative emissions of volatile organic compounds (VOCs), causing much of the air pollution problems in California, could be reduced about 90% using alcohol automotive fuels. However, the blending of methanol with gasoline can lead to higher evaporation of hydrocarbon vapors. A reduction in carbon monoxide emissions can also result, even with alcohol/gasoline blends rather than pure gasoline. Of the two alcohols, ethanol appears superior operationally on environmental grounds. Whereas none of the emission products of ethanol are known to be carcinogenic or mutagenic, formaldehyde is a product of methanol combustion and is a suspected carcinogen. In addition, methanol poses more toxic hazards in use than gasoline.

The projected production costs of ethanol from cellulosic feed stocks have fallen since the early 1980s by a factor of three to a range of estimates $1.05–$1.55/gallon ($0.28–0.41/liter) (Bull, 1991; Wyman *et al.*, 1993; Ahmed, 1994). The DOE Renewable Energy Program has a goal to reduce this cost by about 50% to a level below $0.70/gallon ($0.18/liter) (DOE, 1993, 1994a). (For any attempted comparison to gasoline prices, it should be recognized that this is a production cost, not a market price. Furthermore, it should be remembered that the energy content of a gallon of alcohol is a fraction of that of a gallon of gasoline. With proper engine design, an alcohol will deliver about 80% of the range per gallon achieved with gasoline.) These estimates include the costs of feed stocks up to about $40/dry ton ($44/mt), grown with land productivities up to 10 ton/acre (22mt/ha) dry weight annually. Biochemical processing, of the types described above, accounts for more than half of the overall cost, however. Further reductions in cost are possible through reductions in feed stock costs, by choice of energy crop, or by increasing land productivity to the greatest extent possible. Processing costs are expected to drop, per unit of output, with efficiency of scale for larger plant capacities and with technology improvements (Wyman *et al.*, 1993).

The market prospects for ethanol, in view of these estimates, can be taken as approaching marginal competitiveness with the gasoline prices expected in the 1990s. Currently, an annual alcohol production capacity in the USA around 1 billion gallons can supply less than 2% of the liquid fuel demand. Given the short-run prospects for production costs, similar to those of several other renewable technologies we have considered, little or no penetration could be expected into the automotive fuels markets in the first decade of the 21st century or beyond, in the absence of any other pressures for adoption. Other pressures, however, *are* present for automotive fuels – namely, their environment.

The environmental degradation caused by automotive emissions has reached the acute stage in California and other major regions throughout the USA (EPA, 1989; Gordon, 1991; Krupnick & Portney, 1991). Reduced-emission or zero-emission vehicles are being mandated for the near future by state regulation in California, and other states seem to be ready to follow. Pressures for R&D have heightened and further subsidies for non-polluting fuels or electric vehicles are likely to be legislated. As emphasized in our introduction, automotive fuels is the most dramatic example of an emerging transformation in energy policies, driven by rising environmental concerns. A new perspective is developing that considers that externalities, indirect costs borne by the public (resulting, for example, from pollution, property, or health damage), should be included with internal costs (those directly affecting the market price) when assessing the true costs of various energy sources.

Methanol ("wood alcohol") has been considered as a competitor to ethanol for an alcohol automotive fuel (Moreno & Bailey, 1989; Kohl, 1990; Archer, 1992), either in pure form or as a blend with gasoline. A means of producing methanol from biomass feed stocks, starting with thermochemical gasifiers, is discussed in the following section. Whereas methanol can be produced from biomass feed stocks, recent development efforts in the USA have focused on natural gas (nearly pure methane) as the feed stock (Kohl, ed., 1990; NRC, 1990), looking to more immediate production and marketability for improvement of air quality rather than getting away from use of fossil-fuel feedstocks. As such, this and other means of enhancing the use of alternatives to gasoline are being considered, by some analysts, as "transition" solutions to truly renewable technologies. In the process, however, many of the developments by automobile manufacturers for methanol will be transferable to biomass-fed ethanol, which is considered generally superior from the environmental/

climate point of view.

The 1992–96 Biofuels Program of the US DOE includes projects for both ethanol and methanol (NREL, 1992; DOE, 1993). For ethanol, the emphasis is on the use of cellulosic (non-food crops, wood, and solid waste) feed stocks, with specific goals of production costs at $0.67/gallon ($0.18/liter). Such production would require feed stock at a cost below $35/ton ($38/mt) (dry weight), with annual land productivity at least 10 ton/acre (dry weight) (22mt/ha). Processing plant capacities of nearly 2000 tons (1800 mt) of biomass per day would be needed to achieve mass production economies in conversion to ethanol. Of these goals, land productivity has already been largely achieved, with an increase in proto-type SRWC production from 5–7 ton/acre (11–16 mt/ha) to 10 ton/acre (22 mt/ha) from 1993 to 1998 and similar yields with the energy crops and grasses. There is an overall DOE goal for a cellulosic-based ethanol technology ready for commercial deployment by the year 2000 (Saha &Woodward, 1997). Switchgrass and hybrid poplar feedstocks will be emphasized, but a diversity of other feedstocks will be maintained.

The US R&D program is managed and led by the National Renewable Energy Laboratory (NREL) at Golden, Colorado (one of DOE's national laboratories, formerly named the Solar Energy Research Institute). One major project involves the New Energy Company of Indiana, which is developing the SSF process for production of ethanol from corn fiber as a step toward a wider use of cellulosic feed stocks. Another project, based entirely on non-food feed stocks, is with the Amoco Research Center and is striving for economic ethanol production from wood, grasses, or various wastes. A pilot production operation for cellulosic ethanol has been set up at NREL itself (NREL, 1995). In addition, various other research projects are occuring on smaller scales at NREL and various universities. These are laboratory projects concerning biochemical reactions and alternative pro-cessing techniques. In addition to these R&D activities, the DOE and the USDA are engaged in programs promoting biofuels (USDA, 1992; DOE, 1996b), including alcohols. Figure 2.8 is an example of an operating prototype of an alcohol-fueled automobile.

2.3 Gaseous biofuels

Gaseous fuels, mainly methane, are under consideration to replace natural gas: a fossil fuel. Here the use is for stationary energy production to supply heat for residential and commercial needs and for industrial processing. Conceivably, a biogas could also substitute for natural gas in electric power plants. In order to substitute directly for natural gas, a biomass methane would have to be of "pipeline quality" in heat content (around 1000 BTU/ft^3 in the USA).

Gaseous biofuels may be produced by biochemical reactions, such as anaerobic decomposition, or by thermochemical means, such as pyrolysis or gasification (Brink *et al.*, 1976; Smith & Frank, 1988). The products vary but include methane, hydrogen, and carbon monoxide, as combus-

Figure 2.8
An ethanol-fueled
automobile.(Courtesy
of the Ford Motor
Company.)

tibles. The ultimate gaseous-fuel product is a high-BTU gas, close to pipeline quality, whether it is derived from biomass or coal as the feed stock. A product of lower heating value (500–700/BTU/ft^3) has limited uses for industrial heating or as the input for a liquefaction process.

Biochemical production

The principal biochemical means for methane production is anaerobic digestion. This process has been known and used since the late 19th century but like other energy sources was abandoned in the industrialized world with the advent of cheap fossil fuels. Interest revived with the oil crisis of the 1970s. Methane digesters have gained wide use in LDCs since the 1970s (Aziz, 1991) to the point where millions of the devices are in use in the 1990s (Brower, 1992), fed mostly by animal dung.

The digestion process takes place in three phases: hydrolysis, acid formation, and methanation (Meynell, 1982; Smith & Frank, 1988). The hydrolysis step is the biochemical equivalent of the step that precedes fermentation for alcohols. In this case, bacteria break down the organic molecules of starch and cellulose in the feed stocks (usually organic wastes). This process is followed directly by a conversion to fatty acids (e.g. acetic acid), all taking place within a time span of a day or so. Methanation will follow if the acids are neutralized or diluted (sometimes, simply by addition of more feed stock). Once the methanogenic bacteria are able to function, methane production is completed in 3 to 4 weeks, depending on the temperature in the digester.

The typical digester used in the Third World is simple in construction (Fig. 2.9). It consists of a digestion chamber, an inlet and an outlet for solids, a mixing agitator, and a gas outlet chamber. The digestion chamber can be a metal tank or simply a pit in the ground lined with brick or stone. Inlet and outlet openings may be provided for continuous processing but

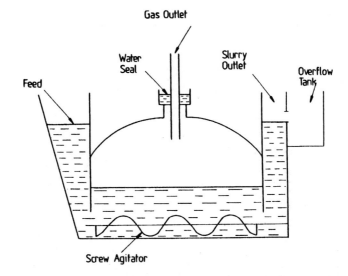

Figure 2.9
A Third World biogas digester. The digester is constructed from masonry. (Reprinted with permission from *Renewable Energies: Sources, Conversion and Application*, P.D. Dunn, Peregrinus Press, London, 1986.)

Figure 2.10
A biomass gasifier in Vermont. (Courtesy of the National Renewable Energy Laboratory; *photo credit*: Glenn Farris, FERCO.)

would not be necessary for batch processing. Some sort of stirring or agitation is required, in the form of rotating paddles or screw mechanisms, no matter what the processing. The gas outlet is, in most of the various digesters throughout these countries, a heavy metal dome that traps the biogas above the liquid slurry and maintains a gas pressure for release through the gas outlet pipe.

These simple digesters, in the methanation phase, can produce daily volumes of gas roughly equal to the volumes of the digesters (typically about 1500 ft^3) and can supply the needs of one family. Larger and more elaborately controlled digesters (Fig. 2.10 for a working example) have also been built and operated in developing countries as well as in industrial countries. In the developing countries, the feed stocks are usually animal dung, whether on the family, village, or even industrial scale. In the industrial countries, methane digesters are more likely to use sewage sludge, but relatively few such operations currently exist. A few farm digesters (estimated at 100 for the USA (Andrejko, 1989)) are operated in the industrialized countries.

The composition of biogas digesters can range from 50 to 65% methane, with the remainder being mainly carbon dioxide, depending on the feed stock and the completeness of the process (Meynell, 1982; Aziz, 1991). This puts this biogas in the class of a medium-BTU gas, since it will not be of pipeline quality unless it is nearly 100% methane. The possible feed stocks include food wastes and raw garbage, including organics other than food wastes, animal dung (all types), and sewage sludge. Gas production volumes for a 50% methane biogas can be in the range of 5–10 ft^3/lb (0.3–0 6 m^3/kg) feed stock, which would be equivalent, of course, to 2.5–5 ft^3/lb (0.16–0.31 m^3/kg) natural gas.

Anaerobic treatment of the organics in industrial waste waters is gaining popularity in the industrialized world (Lettinga & van Haandel, 1993). The types of waste water treated include those from food processing, dairy, brewing, pharmaceutical, paper and pulp, and, even, alcohol production. The "high-rate" anaerobic processes used are more sophisticated than those just described. Typical of these more advanced processes is the "upflow anaerobic sludge blanket" (UASB) method, which achieves a granulation of the waste-water sludge as it moves upward out of a blanket at the bottom of the reactor. This movement through higher zones in the reactor results in the formation of granules and films in which the bacterial growth is enhanced greatly over that occurring immersed in the solid sludge. Expert operation is necessary, however, in start up of the reactors (with seeding bacteria) and to control the process, once progressing, as to temperature, organic composition, and alkalinity.

Another interesting prospect for anaerobic processing is to use it on the waste liquid from ethanol production (called "vinasse") (Lettinga & van Haandel, 1993). The energy content of the methane resulting from anaerobic processing of the vinasse adds 10% more energy from the original feed stock (e.g. sugar cane) to the 38% conversion by fermentation, to make a total of 48% energy conversion. These results have been achieved in the Brazilian "Pro-a'cool" Program.

Anaerobic treatment of waste-water streams not only produces a usable methane fuel but can also be part of water-pollution control measures. Whereas, aerobic decontamination of municipal sewage is the usual approach in the USA and other countries, it requires (electric) energy to operate. Anaerobic decontamination, by comparison, can be a net energy producer. The situation here, regarding net energy production, is similar to that found in MSW disposal (see below). These technologies may be important to municipalities in reducing their costs of disposal, but they will not be major energy resources in themselves on a nationwide scale.

Anaerobic processing of waste waters could, however, play a role elsewhere for biofuels. Water-pollution control, if large-scale alcohol production is adopted, poses major environmental challenges (Lettinga & van Haandel, 1993). For instance, the current production level of alcohol in Brazil (3.5 billion gallons annually) creates water pollution equivalent to that produced by 160 million people (greater than the population of that country). Consequently, anaerobic treatment of alcohol-production waste waters makes sense on several counts. However, it should be recognized that anaerobic processing is only a means of dealing with *organic* pollutants in waste water; other means, such as filtering, must still be used to separate out inorganic pollutants such as heavy metals (Speece, 1985).

Another waste resource for anaerobic methane production is municipal landfills. In this case, decomposition is taking place in an oxygen-deficient atmosphere under the fill dirt and not in a designed reactor. The resulting gas is about 50% methane (medium-BTU content) (EPA, 1992). The landfill gas, which otherwise must be vented and flared for reasons of safety, can be tapped for use as a fuel. Investigation of the extraction of landfill gas took place in the USA during the late 1970s, under the impetus of the oil crisis (IGT, 1978). Elaborate underground piping systems were designed in some prototype operations and processing for pipeline-quality gas was attempted.

It was found in these early projects in the USA that the amount of landfill gas produced in the average landfill was not large enough to be of commercial interest, especially considering the costs of processing and (pipe) transmission to customers. However, pilot projects are still being carried on at various locations around the world (Warmer Campaign, 1991–94). Recent estimates suggest that the capital costs of an anaerobic digestion plant with an annual capacity of 50 000 tons (45 000 mt) would be approximately $20M, which is less than a mass-burn plant of the same capacity. The question remains, perhaps, as to how this methane could be best used: directly as a fuel for heat generation or for conversion to other forms of energy, such as electricity.

Methane can be used as the input to appropriately designed fuel cells. Recently, the US Environmental Protection Agency (EPA) has promoted fuel-cell power plants using the landfill methane on site (EPA, 1992). Combustion turbines and methane-fueled engines driving conventional electric generators are also being used on landfill sites. These generating units are typically of moderate output capacity (less than 1 MW) and are not of major commercial importance for nearby electric utilities. The primary motivation of the EPA in these projects is to control methane emissions from the landfills. Again, like the other trash-to-energy schemes, revenues from the electricity generated at these landfills would partially offset the costs of disposal for the municipalities but would not result in a net profit for the operation.

As many as 5000 landfills in the USA could be tapped profitably for gas extraction, yielding a possible aggregate 0.2 QUAD annually (0.2 QUAD is barely measurable on the national energy consumption of about 80 QUAD or 84 EJ). As of the mid-1990s, there are over 200 landfill gas projects operating in the USA, but their products (both gas and electricity) are often undercut by the markets (Williams & Bateman, 1995). The EPA, however, is promoting these operations out of the need to control methane emissions. Landfill gas plants are also operating in Austria, France, Finland, Great Britain, and Tahiti, operating with annual waste-handling capacities of tens of thousands of tons (Warmer Campaign, 1991–94).

Finally, feed stocks other than wastes have been proposed for anaerobic methane production (Smith & Frank, 1988). These other feed stocks include not only the energy crops grown for alcohol fermentation but also aquatic plants such as water hyacinth. However, only by resorting to mass production of energy crops, such as considered for the alcohol fuels, could biomass methane be produced on a scale significant to the present consumption of (fossil-fuel) natural gas in the USA. If a national policy for developing energy-crop production is initiated here, decisions may then have to be made at the initial stages concerning the allocation of feedstocks to alcohol or methane production. The two fuels are not, of course, direct substitutes for one another: one is liquid and the other is a gas. Nonetheless, the two markets will be those competing against petroleum (mainly for gasoline) and natural gas, both of which are major suppliers of the economy and both of which have significant cross-dependencies. Once production and market penetration has developed, then, of course, market demand could direct the allocations.

A major factor in the development of major biomass methane production will be the processing costs for high-rate anaerobic methanation. Estimates to date vary widely, with the US DOE/Gas Research Institute programs stating production costs around $5/MBTU ($4.70/GJ) (Andrejko, 1989), while the international United Nations Conference on Environment and Development (at Rio de Janeiro, 1992) (Lettinga & van Haandel, 1993) has optimistically estimated less than $1/MBTU ($1.05/GJ), presumably based on experience with the UASB reactors. Such figures again reflect production costs only, not sales prices, and do not even include transmission and distribution. The lower estimated production cost is clearly the range necessary for competition with natural gas, being in the same range as current wellhead prices for natural gas.

Thermochemical production

Biomass feed stocks, such as wood and herbaceous crop residues, may be processed thermochemically to produce gaseous and liquid fuels (Miles & Miles, 1989; Bull, 1991; Williams & Larson, 1993). A major thermochemical process is *pyrolysis*, in which the complex organic units that comprise cellulose, lignin, and carbohydrates are broken down into more elemental compounds and forms of carbon, hydrogen, and oxygen.

Succinctly defined, pyrolysis is the "destructive distillation" of organic material, when heated to temperatures above 600 °C, in an atmosphere that is too deficient in oxygen for combustion to take place (Dunn, 1986; Miles & Miles, 1989). The conditions for pyrolysis have been created in wood kilns for production of charcoal going back into history and are used today in the underdeveloped countries, but the gaseous products have not previously been utilized in these more primitive devices. More recently, however, kilns used as gasifiers have come into use in the developing countries (Mukunda *et al.*, 1993).

The processes and gaseous products of pyrolysis with biomass feed stocks are quite similar to those with coal as the feed stock, where coke is the solid product and "town gas" or "producer gas" are the possible gaseous outputs (Reed, 1993; Dunn, 1986). The products of pyrolysis vary widely with the temperatures applied to the organics and with the rate at which the temperature is raised above 600 °C. With a wood feedstock and raising the temperature gradually from 600 °C to around 1000 °C, charcoal (with a heating value comparable to coal: 22–28 MBTU/ton or 26–33 GJ/mt) and a low- to medium-BTU (100-400 BTU/SCF or 3728–14910 GJ/m^3) gas are the results. This gaseous product is composed of combustibles – carbon monoxide, hydrogen, and small amounts of methane – plus non-combustible carbon dioxide and water vapor. Two other thermochemical reactions can also take place in biomass processing owing to the water content of the feedstock, namely the "steam-carbon" reaction and the "hydrogasification reaction". These reactions also are present in coal gasifiers when steam is intentionally introduced into the reactor and can lead to higher concentrations of methane.

It has been asserted that any pyrolysis, gasification, or combustion of the products of a wood feedstock can be more energetically efficient for a thermal power plant than direct combustion of wood, adequate even for IPH requirements (Brink, et. al. 1976). However a rapid but limited increase in temperature results in "flash pyrolysis", leaving decomposition of the biomass compounds incomplete and yielding hydrocarbon gases. If this gas is then condensed, a hydrocarbon liquid called "biocrude" can be obtained. With further treatment, this product holds promise as a synthetic alternative to petroleum, but more development is needed.

Another class of thermochemical processes is (high-temperature) gasification. Like pyrolysis, biomass gasifiers operate on principles similar to coal gasifiers which have purely gaseous fuel products (Miles & Miles, 1989; Williams & Larson, 1993). A well-known example is the Lurgi process, originally invented for coal but which can also be used with biomass feedstocks. The steam–carbon and water–gas reactions, familiar historically from coal processing, take place in a reactor in various temperature zones as the processed material falls through the reactor from the top or side (Fig. 2.11). A small bubbling-bed gasifier (Fig. 2.11b) has been successfully operated on a pilot basis with biomass feedstock by the (gas industry's) Institute of Gas Technology (IGT).

The reactivities of biomass feed stocks are higher than those for coal in the Lurgi processing, resulting in more complete reactions and shorter processing times. The products of the various thermochemical reactions vary widely with the feed stocks and reactor conditions and are still the subject of research (DOE, 1993, 1994c). For example, the role of the lignin components of woods and crops in the process must be better understood in order to make best use of all organic components. Slagging (accumulation of carbon residues) and polluted effluents, problems that had been encountered in earlier coal-derived synfuel programs (Williams & Larson,

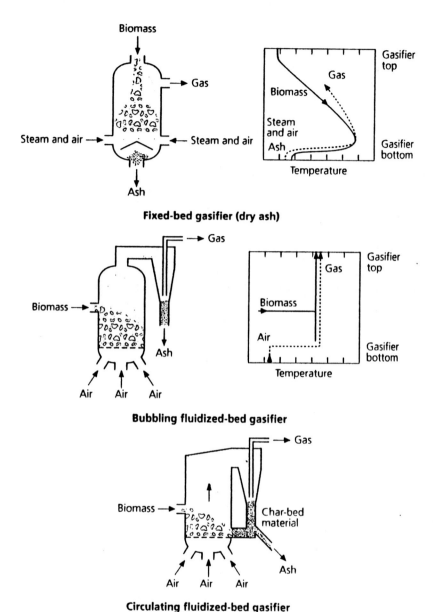

Figure 2.11
Operating principles
and temperature profile
for a fixed-bed gasifier
and a bubbling
fluidized-bed gasfier.
The operating principle
for a circulating
fluidized-bed gasifier is
also shown.
(Reprinted with
permission from
Johansson, T.B. *et al.*,
(eds.) (1993).
*Renewable Energy –
Sources for Fuels and
Electricity*, Island Press,
Washington, DC and
Covello, CA.)

1993), are being addressed. Successes have been reported on "hot-gas
cleanup" (DOE, 1994c) and solutions are being sought on hot-gas conden-
sation using temperature control of the reactions. In this and other research concerns, we can see parallels with those for the alcohol fuels.

Efforts were started early in the 1970's energy-crisis period for development of biomass gasifiers and pyrolytic oils. In the USA, research and
initial development, sponsored by the DOE (and its predecessor ERDA),
was carried on with notable progress into the late 1980s to establish a
working technology to produce a biogas product at near competitive costs.

Innovative integrated co-generation systems combining gasifiers with gas turbines were also proposed, again adapting earlier coal-based schemes to biomass feedstocks. These demonstration projects in the USA focused on wood as the feed stock. Biomass gasifier projects have also been started in Finland and Sweden using wood feedstocks (Patterson, 1994). The THER-MIE program, under joint sponsorship of the countries of the EU, initiated a transition from coal feedstocks to biomass for integrated gasifier installations.

In 1986, federal funding in the USA was cut and final development was left to private interests such as forest-products companies and the gas industry. This, of course, followed the pattern of R&D programs for the other renewable technologies already discussed. Some R&D efforts have since been sponsored by the gas industry (IGT) and private foundations (Williams & Larson, 1993). In 1990, US federal R&D funding was resumed, this time for BIG/GT (biomass-integrated gasifier/gas turbine) systems. The large-scale prototype operation will be run by IGT, benefiting from their experience to date with gasification technology (see more on this below). Also, a pilot BIG/GT operation, using wood as the feedstock, is under consideration by electric utilities in Vermont, with possible federal (DOE, EPA and USAID) cosponsorship.

Many of the simple gasifiers installed in the USA during the earlier DOE program are still operating, although several have been abandoned (Andrejko, 1989). All in all, there were over 3000 biomass combustion systems in the USA in 1989 (Miles & Miles, 1983), of which 35 were gasifiers installed during the DOE program (Antal, 1985). These plants operated, for the most part, in the southeastern USA and were fed by wood. There are over a dozen gasifier/pyrolysis systems marketed at the end of the 1990s, producing low-BTU gas. Thermal gasifiers are being developed and are in operation in both developed countries (Belgium, France, Finland, Italy, and Sweden) and in developing countries (Brazil, India, Indonesia, and Thailand) (Antal, 1983, 1985; Mukunda et al., 1993).

These biogas products, being low- to medium- BTU heat content, are not substitutes for natural gas of "pipeline" quality. As such, their potential markets are limited to industrial and agricultural uses in the industrialized countries, unless there are specific application technology developments for their use. Such developments might be for biogas-fueled engines (Dunn, 1996), a "synthesis gas" for further processing into a high-BTU gas, as the gaseous input for "indirect liquefaction" of (hydrocarbon) petroleum substitutes, or as the synthesis-gas input to methanol production (see below). In order for gasification to be economic on a mass scale, however, production costs would have to be reduced below the previous $8/MBTU ($7.60/GJ) level (Brower, 1992). (A production cost below $3.50/MBTU would be required to be competitive in the USA with natural gas in the medium-term future.)

In the developing world, biogasifiers have been used with success, particularly for agricultural applications. Using either wood or charcoal as feed stocks, the low-BTU biogas can be used in a spark-ignited tractor engine, with the gas stored in a gas bag on top of the vehicle (Dunn, 1986).

The gas can also be used for heat in crop drying, cooking, and food processing. Small biogasifiers are one of several technologies, along with solar and wind, that have already made significant gains for rural improvements in the underdeveloped world.

Higher technology uses for biogasification are being tried in the developed countries (DOE, 1991; Williams & Larson, 1993; Bain & Jones, 1993). As an example, the lower-BTU output of biomass gasifiers can be used in a high-efficiency electricity-generation scheme called combined-cycle (CC) generation. In any CC system, electricity is produced by two generators: the first one driven by a combustion turbine and the second by a steam turbine. The (very hot) exhaust from the combustion turbine is fed to a heat exchanger, which generates steam for the steam turbine. The overall efficiency of a CC system can be made to approach 45%, compared with a limit of about 35% for a conventional (single-cycle) steam-generation plant simply burning coal or wood. A CC operation requires a gaseous or liquid fuel input, which is natural gas or a petroleum derivative (like jet fuel) in conventional usage. If a solid-fuel resource, such as coal or wood, is used, it must be converted into a gaseous or liquid form.

Successful CC operation, using a coal-derived synthetic gas as the fuel for the combustion turbine, was achieved in the Coolwater Project in the 1980s. The Coolwater Project was an "integrated" gasifier/CC operation. In an integrated operation, the output of an on-site gasifier is the fuel input for the gas turbine of the CC system. When coal is the feedstock for the integrated gasifier, as it was for the Coolwater Project, the operation has been called a coal-integrated gasifier/gas turbine (CIG/GT) system. When biomass of any sort is the feedstock, the system is the BIG/GT system mentioned above (Fig. 2.12). Such systems are planned for prototype operation, but there is a need to develop a gas turbine that operates on a low-BTU gas (Overend, 1998). For example, a CC–biomass project is being pursued in the DOE Biomass Power Program (DOE, 1991; Bain & Jones, 1993), in cooperation with the IGT and the Natural Energy Institute in Hawaii. Wood and energy crop (e.g. sugar cane) feed stocks will be used in a prototype 4 MW$_e$ plant processing 100 tons per day of biomass in a fluid-bed gasifier, with "hot-gas cleanup" of resins and particulates of the raw gasifier output (these otherwise degrade turbine performance). A BIG/GT project is also contemplated in Brazil, using wood feedstocks from specially planted forests (Williams & Larson, 1993). No results of technical performance or production costs from such projects are available at the time of writing.

However, preliminary estimates of investment required for a BIG/GT plant suggest unit (electric output) capacity costs at a surprisingly low $900/kW$_e$ (for plants over 50 MW$_e$ capacity), including the added 20% of investment for energy-crop plantation land to supply the feedstock. Dedicated feedstock supply systems (DFSS) have been proposed for these plants (Overend, 1998). Further, the complete economics of BIG systems have been projected using data derived from CIG systems. By assuming that both capital and operating costs will be lower for biomass processing than coal, the generation cost of electricity was optimistically estimated to

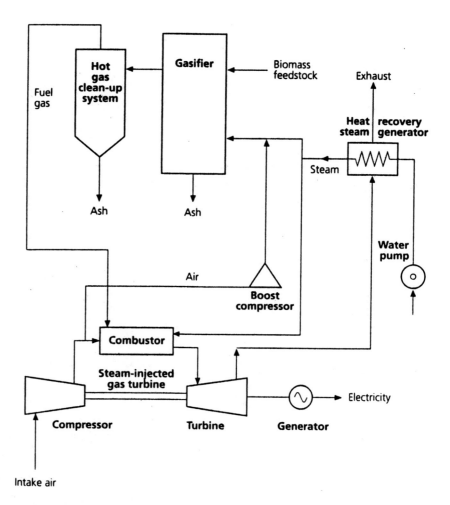

Figure 2.12
A biomass-integrated gasifier (BIG) cycle.(Reprinted with permission from Johansson, T.B. *et al.*, (eds.) (1993) *Renewable Energy – Sources for Fuels and Electricity*, Island Press, Washington, DC and Covello, CA.)

be in the competitive range of $0.05–0.06/kW-hr, even with an assumed high feedstock cost of $60/ton ($66/mt) for wood.

Further development, prototype operations and refinements are planned in the DOE Biomass Power Program, which was initiated in 1990 to demonstrate operational feasibility (Bain & Jones, 1993; Basin & Kirschner, 1998). In the mid-1990s, the program was funded at a level of about $69M per year (DOE, 1995). The BIG/GT technology has the technical potential to supply (electric) energy on a national scale; however, market competitiveness must still be established. Market competitiveness could be proven at a later stage, if development is encouraged with the initial help of incentives such as PURPA and investment credits, as was found necessary for wind-generated electricity. In order to approach competitive operation, feedstock production costs most probably will have to be held to the limits mentioned here earlier, the above optimistic estimates not withstanding.

Another possible route for biogas conversion is to use the outputs of biomass gasifiers as a "synthesis-gas" input to (liquid) methanol production (DOE, 1993; Wyman, *et al.*). Again, such process steps are similar to those previously used with coal gasification prototypes, with the gasifier outputs from biomass feedstocks of hydrogen (30–35%) and carbon monoxide (20–45%), and with smaller amounts of methane and carbon dioxide. The gasifier reactions, which include pyrolysis, (partial) oxidation, and the (steam) water-shift reactions, produce an output with a ratio of hydrogen to carbon monoxide less than one; this ratio must be raised in order to match the hydrogen to carbon monoxide proportions of the methanol synthesis reaction more closely. The carbon dioxide content of the synthesis gas is removed prior to its input and the gases are cleaned of particulates. The methanol-producing reaction itself takes place between the hydrogen and the carbon monoxide, at temperatures over 260 °C (500 °F) in the presence of a catalyst. A conversion of 20–25% is possible per reaction. Prototype production has not yet been attempted for this so-called "indirect" synthesis of methanol, but estimates of production costs have again been made based on experience with coal gasifiers. These costs, covering capital investment, feed stock and operation have been in the range $7.50–18.00/MBTU ($7.10–17.00/GJ), considerably above the range for a competitive automotive fuel ($5–6/MBTU). Further development and experience are needed before operating feasibility and reliable cost figures can be determined for possible commercial production.

Still another route to upgrading the low to medium-BTU output of gasifiers is to convert the mixture mostly to hydrogen (Ogden & Nistch, 1993) rather than to methane. The conversion is accomplished using steam, first with a "steam reforming" of the methane in the mixture to create (more) carbon monoxide and hydrogen. Then, first at a high temperature and then at a lower temperature, the steam reacts with the carbon monoxide to form hydrogen and carbon dioxide. The carbon dioxide is subsequently removed leaving nearly pure hydrogen gas. (Commercial processes for production of hydrogen currently use steam reforming of natural gas, which is nearly pure methane.) The prospects for production of hydrogen fuels will be discussed in Chapter 9. However, questions can be raised at this point as to the value of this indirect approach to the production of hydrogen when direct methods are deemed possible. This is especially pertinent when it is realized that biofuels offer in themselves possible sustainable mass energy sources.

Biomass pyrolytic/gasification conversion of the various types mentioned is still in the R&D stage (DOE, 1993), with commercialization uncertain in the medium-term time frame. For the present state of development, capital costs for the various thermal gaseous and liquid fuel processes have been found in most cases to be higher, per unit capacity of energy delivered, than the production facilities for conventional fossil fuels (oil and natural gas) that they would have to displace. Feedstock costs must also be lowered to reduce the variable costs of production. The goal of the DOE Biofuels Program is to achieve fuel unit costs in a competitive range below $2/MBTU ($1.90/GJ) for several of these processes.

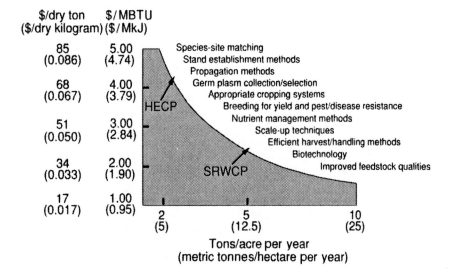

Figure 2.13
Biomass feedstock costs versus land productivity. HECP, herbaceous energy crop production; SRWCP, short-rotation woody crop production. (Courtesy of the National Renewable Energy Laboratory.)

Table 2.1. *Production costs of energy crops for typical sites in the Midwest of the USA*

Cost component	Hybrid poplar Minnesota	Switchgrass Nebraska
Establishment ($/acre)		
Herbicides	6.91	0.26
Fertilizer/liming	3.31	2.36
Machinery	2.64	1.27
Planting	5.01	4.65
Maintenance ($/acre)		
Pesticides	11.84	0.00
Fertilizer	5.52	39.90
Land rent and taxes	75.48	57.50
Managerial	16.57	16.57
Harvesting ($/acre)	128.63	45.98
Total cost ($/acre)	255.92	168.48
Gross yield (tons/acre)	7.00	5.00
Net yield (tons/acre)	5.95	4.25
Total production and harvesting cost ($/ton)	43.01	39.64
Transportation and baling ($/ton)	7.97	9.69
Total cost ($/ton)	50.98	49.33

Source: with permission from Brower, M.C. *et al.* (1993). *Powering the Midwest – Renewable Electricity for the Economy and the Environment.* Cambridge MA: Union of Concerned Scientists.

Table 2.2. *Key research strategies for energy crops*

Identify and characterize, through Geographic Information System (GIS) evaluations and resource analysis, subregions and locations within the North Central and South/Southeastern USA that offer the greatest opportunity for environmentally beneficial, cost-effective energy crop production

Concentrate crop development and cropping systems research on one woody species (poplars) and one herbaceous species (switchgrass) while continuing to screen other species and systems for their adaptability to varied site, climatic, and pest situations

Continue hybrid poplar crop development through breeding, selection, and biotechnology to identify clonal resistance to septoria canker and leaf rust (serious diseases known to cause significant yield reductions in the North Central and Pacific Northwest regions) and resistance to drought stress conditions to ensure long-term crop sustainability

Improve switchgrass yield potential by concentrating on the near-term goal of selecting the best available varieties for the South/Southeast and the long-term goal of breeding improved varieties for a broad range of sites and conditions

Develop environmentally beneficial and cost-effective methods of feedstock production by incorporating superior clones and varieties into field trials and testing methods of establishment, nutrient and pest management, and harvest

Develop and use national, regional, and farm-scale models to define the economics necessary for obtaining landowner interest in energy crop production and for describing the environmental effects likely to occur as a result of changing land-use patterns

Coordinate environmental evaluation with industries, environmental groups, and a variety of local, state, and federal agencies to obtain consensus on critical issues, experimental procedures, and analytical approaches. Collect preliminary data on the effects of energy crops on plant diversity, wildlife habitat, water quality, nutrient utilization, and erosion

Source: National Renewable Energy Laboratory, Golden, CO.

2.4 Conclusions

A universal measure of the market competitiveness of any of these fuels is their cost per unit of heating value ($/MBTU or $/GJ). The goal given above for gaseous fuels of $2/MBTU ($1.89/GJ) can also be applied to the liquid fuels and is translatable to the $/gallon measure customary for liquid transportation fuels. This low cost can be achieved for either fuel type only with feedstocks costs that have been reduced to below $35 per (dry) ton ($38/mt) (dry weight). Such a cost for the crops can, in turn, be attained with energy-crop land productivity approaching 10 ton/acre (22 mt/ha) annually (Fig. 2.13).

As discussed above, energy-crop production costs are incurred for fertilizers, herbicides, planting, irrigation, harvesting, transportation, machinery investments, land rent and taxes. Earlier estimates of costs for both wood and herbaceous crops in the Midwest (Brower, *et al.*, 1993) suggested a level of $50/ton (Table 2.1), which by the DOE relationships (Fig. 2.13) would imply a feedstock unit energy cost of about $3/MBTU ($2.85/GJ). Even more recently, it has been judged that $30/ton ($33/mt) feedstock might be possible (DOE, 1994b), suggesting energy costs below $2/MBTU ($1.90/GJ) for the feedstock. However, since this is the energy cost *only* of the feedstock not yet processed into the form of fuel, the resulting unit costs of the fuels would be considerably higher and, therefore, likely still to be in the non-competitive range. Nonetheless, the DOE goals for feedstocks are being pursued following the research strategies listed in Table 2.2, while implementing the improved cropping methods shown above the curve in Fig. 2.13, in a straightforward program intended to accomplish the projected goals for feedstocks. Similar R&D strategies are being pursued in the EU (EUREC, 1996).

The reduction of processing costs presents an additional challenge for biofuels for which the outcome is still uncertain. For each of the liquid or gaseous categories, a process must be found where reductions in cost can be made in order to reach competitive levels. For the liquid (alcohol) fuels, with transportation as the major use, production costs must be lowered so that the product has a unit cost below $6/MBTU ($5.70/GJ), as an overall cost of production including feedstocks, in order for it to be competitive with current prices of petroleum products. For the gaseous fuels to compete with natural gas and coal for stationary uses, the unit cost overall of the product must be below $2/MBTU ($1.89/GJ). These cost requirements ultimately will depend on market prices for the fossil fuels. Basing these assessments on current market price levels seems reasonable for the medium-term future, however.

Until the costs of biomass fuels become competitive with fossil fuels, significant penetration of the fuels markets cannot be expected to take place, all other things being equal. The absence of any additional pressures for adoption of alternative sources and the level of fossil-fuel market prices may continue to remove incentives for these alternatives, repeating the history of alternative fuels. Nonetheless, there are environmental concerns, such as smog in California, that have begun to generate public pressures for alternatives to pollution from automobile gasoline and diesel fuels and there may be a possible *willingness to pay* higher prices for cleaner automotive fuels (Krupnick & Portney, 1991; Cassedy, 1993). In addition, increased employment prospects resulting from a large new biofuels industry might also be a powerful incentive for initiating the program (Grassi, 1996). The Brazilian ethanol program created about 400000 jobs in the agricultural and industrial sectors (Demetrius, 1990). This program is only a fraction of the size of the industry contemplated in the USA.

Gaseous fuels are mostly used for stationary purposes, an area where environmental awareness has not as yet focused, except for sources of acid rain. Biomass fuels for stationary uses – mainly power plants and industrial processing – could be presented as an alternative to coal, but this would most likely have to be introduced as a later development in the evolution of the pollution-control technologies beyond current efforts to clean ("scrub") the effluents of coal. Such a step would move these

(stationary) technologies toward true sustainability by providing low-pol-
lution sources rather than merely controlling the emissions from inherently
dirty sources (Beder, 1994). The potential for conversion from coal to
biofuels for electric power plants, in fact, has been studied in some detail
for the Midwest of the USA, a region that has been identified as a major
source of acid rain (Brower *et al.*, 1997). This industry using gaseous
biofuels, if viable, would also be very large and again would create significant new employment, just as was anticipated for the coal-based synfuels
of the 1970s.

The other pressures for biomass alternatives that might possibly
countermand the customary market forces on fuels could come from
policies attempting to offset the threat of global warming. However, such
policies are, by the long-range nature of climate change, not likely to reach
the acute stage that regional air pollution has and, therefore, are not likely
to be as encompassing in the absence of strong political leadership. Consequently, in the absence of strong public action and without a change in
external factors, the development of biofuels in the near to medium term
will be governed by the prices occurring in the fuels markets.

If a massive biofuels program is ever adopted as public policy, on
grounds of environmental and climate protection, it must be consistent
with such goals throughout. Sustainable agriculture and silvaculture practices will have to be used in production of feedstocks, and ecological
impacts, such as biodiversity, should be considered (Alterieri, 1995).
Fossil-fuel usage in feedstock production must be minimized and, of
course, there has to be an output/input energy ratio much higher than
unity in all feedstock production. Environmentally accepted and occupa-
tionally safe industry standards must be adhered to in biofuels processing
operations. Goals could be set for the biofuels industry, similar to those
recently enacted for the US food industry, to preserve soils, water quality,
and wildlife in feedstock agriculture (discussed in the journal *Science*
(Anon., 1995)). In addition, health and safety standards for workers in
both the agricultural and industrial phases of the industry must be maintained.

Meeting such goals is no small task since the basic motivation for a
biomass program is to produce feedstocks on the huge scales sufficient for
their energy to displace the present massive use of fossil-fuel energy. This
will require mass production of energy crops and wood on scales comparable to or more than those presently used for food crops and timber
harvesting. The record worldwide to date for impacts of mass-production
agriculture and timbering is plainly evident, with soil depletion, erosion,
pesticide contamination, water pollution, and water resource depletion in
both developed and underdeveloped countries. Present-day food agriculture, under the pressures of high commercial demand in the industrial
countries and growing populations in the developing nations, is employing
fertilizer, pesticide, and energy (fossil-fuel) inputs to boost productivity
that are deemed unsustainable by knowledgeable observers (Pimental &
Giampietro, 1994; Perrings *et al.*, 1995; Brown & Flavin, 1996). Present-
day timbering and tree plantations by commercial interests in the Third
World are employing methods that are degrading soil fertility, causing soil

loss through erosion, destroying natural habitats, and disrupting the social fabrics of whole communities (Boyle & Boyle, 1994; Abramovitz, 1998).

While only a few of these large-scale operations are, at present, for fuels or fuel feedstocks, production methods and low-cost objectives parallel those proposed for energy biomass feedstocks. Field crops, such as corn, provide examples where large energy and fertilizer inputs are used to boost yields. (Some corn crops are even now used as feedstocks for ethanol production in a US subsidized program that is criticized because of these heavy non-sustainable inputs (Schippler & Sperling, 1994).) Non-timber tree plantations, growing selected fast-growing species to get high production for paper pulp feedstocks (Mattoon, 1998), have precisely the same low-cost objectives as the cellulosic feedstock production we have reviewed here for biofuel production. Critics have contended that these large-scale plantations are monoculture agroforestry, promoting non-native species such as eucalyptus that are threatening the fertility of soils, hydrologic basins, and biodiversity of plant and animal species in wide tropical areas (Brown *et al.*, 1997). Biodiversity is also of concern for Third-World agricultural development programs (Pagiola *et al.*, 1998).

While such threats have not been ignored in the R&D efforts in the USA and Europe for biofuels (Miles & Miles, 1989; van der Bijl, 1996; Kort *et al.*,1998), a note of caution may still be in order. Environmental research in the USA is being carried out on soil and water conservation, water quality, biodiversity, and wildlife habitats in cropping and tree plantations for energy biomass. Erosion and soil loss are of particular concern (Kort *et al.*, 1998). Preliminary results suggest that crops such as switchgrass can contribute to reducing runoff and sediments (nitrogen and phosphorus) movement to watersheds compared with the effect of conventional crops such as wheat and alfalfa. Other early results indicate environmental benefits from energy crops, such as buffer strips of switchgrass to reduce erosion and to cleanse ground waters.

Hardwood tree plantings have also been observed to increase the diversity of resident bird and small mammal populations over those for the same area in conventional row crops. Recent observations indicate that new-growth forests may have an even greater capacity for carbon dioxide sequestration than previously believed (Anon., 1997a). Finally, there are reports that monoculture plantations will not necessarily be subject to attack by disease and pests solely as a result of the susceptibility of a single species, given appropriate selection of site and proper land-clearing methods (Hall *et al.*, 1993; Hall, 1994). In addition, efforts are being considered to preserve biodiversity through seed banks and planting in botanical gardens (Brown et.al, 1997).

These results are all preliminary. Much more is required in research and carefully conducted pilot operations of energy-crop agriculture and silvaculture before large-scale production of feedstocks can be considered (El Bassam, 1998; Tuscan, 1998). Impacts, unforeseen for lack of thorough exploration, on the scales here contemplated would be unconscionable. All possible impacts should be examined including resource use, soil preservation, energy balances, pollutant emissions, greenhouse-gas balances, and (natural) biodiversity, as well as economic and social impacts (Alterieri,

1995). Better anticipation of the possible effects of monocultures and selective breeding of the energy crops themselves is desirable too, in order to better guard against unforeseen susceptibility to particular diseases and pests.

Research on species improvement to date appears to have been mostly confined to traditional breeding techniques to improve genetic traits (DOE, 1994c). Transgenic techniques, splicing in entirely new genes into the plant genome, have been considered as an area of research for energy crops (El Bassam, 1998; Tuscan, 1998). These techniques have been explored extensively in (food) agricultural research, sometimes with unpredictable outcomes according to critics (Rissler & Mellon, 1996). These would include inadvertent breeding of wild strains that have the potential to create widespread weed pests and reduce biodiversity. Given the likely pressures to produce superstrain crops or trees for energy feedstocks, some caution might be wise in view of these possible risks (Anon., 1998).

The final consideration for the impacts of biofuels is for *processing*, keeping in mind again the massive operations that are being proposed here. These biofuel operations would have to be on the scale of the present oil and natural gas industries in order to produce liquid and gaseous biofuels to displace significant fractions of the outputs of those industries. Processing operations on such a large scale were previously considered for the Syn Fuel programs of the 1970s, using coal as the feedstock (Stobough & Yergin, 1979; Cassedy & Grossman, 1990). These biomass programs would parallel the coal processing operations of the Syn Fuel Program in several aspects, as we have already discussed. While coal processing did not progress beyond pilot operations for mass operations (before succumbing to financial difficulties), a good deal of experience in processing organic substances was gained. It is to be hoped that experience from these coal projects, concerning environmental and occupational hazards might also be useful in planning for mass processing for biofuels since processing for gaseous biofuels parallels that for coal: not only are the end products (e.g. methane) the same but many of the by-products and emissions are also the same. While biomass will not have the sulfur compound emissions of coal, such as sulfur dioxide, it can still have emissions of nitrogen oxides, resulting from combustion or other thermal reactions in air (Wyman *et al.*, 1993). It will also produce tars and aromatic gases from gasification reactions and have emissions carrying particulates, carbon monoxide, and possibly trace elements (Ranney & Cushman, 1991). While there are reports that scrubbing, gas cleaning, or optimized operation (e.g. IGCC) can reduce such emissions (Beenackers & Miniatis, 1996), further work will undoubtedly be required for large-scale operations to meet sustainable standards.

Occupational health and safety in operations and processing was a concern in the coal Syn Fuel Program, as far as it went (DOE, 1981) and must be considered carefully for the biomass program. In the biomass case, even before the processing stage, there are possible dangers of explosive mixtures in the storage of feedstocks (Wilen & Rautalin, 1996). In gaseous fuel processing, the generation of toxic gases and carcinogens was a concern in the coal programs and is also a possibility in the biomass program. These include toxic hydrogen cyanide, ammonia, and carbon monoxide gases and carcinogenic liquids such as phenols, formaldehyde, and tars.

For large-scale processing of liquid biofuels, such as alcohols, liquid wastes and gaseous emissions must be handled for environmental, health, and safety protection. Liquid wastes would possibly include acids, alkalis, phenols, and various of the enzymes used, which must be considered as water pollution requiring processing. In addition, high levels of dissolved organics, including sugars, carbohydrates, yeasts, and various cellulosics in waste waters can lead to conditions of high biological oxygen demand (BOD) (e.g.. the "red tide") in surrounding waters if these compounds are not removed before discharge of waste waters.

While the prospects for sustainability of biofuels is very positive overall – certainly when compared with fossil fuels (Angulo *et al.*, 1998) – a note of caution has been inserted here because of the huge scale involved. The overall cycle of production, starting from the growing of feedstocks and going through the conversion processes to the stage of fuel, should be evaluated for environmental, occupational, and social impacts before any commitment to large-scale operations are put into motion. A promising approach for such comprehensive impact evaluation is "life cycle assessment" (LCA) (Curran, 1996; Kaltschmitt *et al.*, 1996), which is discussed for more general application in the final summaries at the end of Chapter 1.

References

Abramovitz, J.M. (1998). *Sustaining the World's Forests, State of the World 1998*, pp. 21–40. New York: Norton.

Adler, M.J. & Schwengels, P. (1991). Issues in developing a biomass energy strategy for reducing greenhouse gas emissions. In: *Solar World Congress*, Vol. 1, Part II, eds. Adler, M.E. *et al.*, pp. 759–765. Oxford: Pergamon Press.

Ahmed, K. (1994). Renewable energy technologies – a review of the status and costs of selected technologies. *Energy Series, Technical Paper No. 240*. Washington, DC: The World Bank.

Alterieri, M.A. (1995). *Agroecology – The Science of Sustainable Agriculture*, Boulder, CO: Westview Press.

Andrejko, D.A. (ed.) (1989). *Assessment of Solar Energy Technologies*. Boulder, CO: American Solar Energy Society.

Angulo, R.M.S., Linares, R. & Teal, J. (1998). Assessment of the externalities of biomass energy and a comparison with coal. *Biomass and Biomass Energy*, 14, 469–478.

Anon. (1995). Sustainable agriculture and the (1995) Farm Bill (Editorial). *Science*, 267, 943.

Anon. (1997a). Resurgent forests can be greenhouse sponges (Research News). *Science*, 277, 315–316.

Anon. (1997b). Reseeding the green revolution (Special News Report: World Food Prospects). *Science*, 277, 1038–1043.

Anon. (1998). Agricultural biotech faces backlash in Europe (News Focus). *Science*, 281, 768–770.

Antal, M.J. (1983). Biomass pyrolysis: a review of the literature. In: *Advances in Solar Energy*, Part I, *Carbohydrate pyrolysis*, Vol. 1, eds. Boer, K.W. & Duffie, J.A., pp. 61–107, Boulder, CO: American Solar Energy Society.

Antal, M.J. (1985). Biomass pyrolysis: a review of the literature. In: *Advances in Solar Energy*, Part II, *Lignocellulose pyrolysis* , Vol. 2, eds. Boer, K.W. & Duffie, J.A., pp. 175–256. Boulder, CO: American Solar Energy Society.

Archer, L.J. (1992). *Exhausting Our Options: Fuel Efficient Cars and the Environment*. Oxford: Oxford Institute for Energy Studies, Parchment (Oxford) Ltd.

Asplund, B. (1996). The European bioenergy in forestry. In: *Biomass for Energy and Environment*, Vol. 1, eds. Chartier, P., Ferrero, G.L., Henius, U.M., Hultberg, S. & Sachau, J., pp. 78–83. London: Pergamon, Elsevier.

Aziz, M.A. (1991). Biogas: an assessment of potentials, technologies and utilization in Asia. In: *Energy & Environmental Progress*, Vol. B, ed. Veziroglu, T.N. Commack, NY: Nova Science.

Badin, J. & Kirschner, J. (1998). Biomass greens US power production *Renewable Energy World*, 1, 40–45.

Bain, R. L. & Jones, J. (1993). Renewable electricity from biomass. *Solar Today*, May/June, 21–23.

Bain, R. L. & Overend, R.P. (1992). Biomass electric technologies status and future development. In: *Advances in Solar Energy*, Vol.7, Ch. 1, ed. Boer, K.W. Boulder, CO: American Solar Energy Society.

Beder, S. (1994). The role of technology in sustainable development. *IEEE Technology & Society Magazine*, 13 (Winter), 14–19.

Beenackers, A.A.C.M. & Miniatis, K. (1996). Gasification technologies for heat and power from biomass. In: *Biomass for Energy and Environment*, Vol. 1, eds. Chartier, P., Ferrero, G.L., Henius, U.M., Hultberg, S. & Sachau, J., pp. 228–259. London: Pergamon, Elsevier..

Boyle, T.J.B. & Boyle, G.E.B. (1994). *Biodiversity, Temperate Ecosystems and Global Change*. Berlin: Springer-Verlag.

Brink, D.L. Charley, J.A. Faltico, G.W. & Thomas, J.F. (1976). The pyrolysis-gasification–combustion process energy considerations and overall processing. In: *Thermal Uses and Properties of Carbohydrates and Lignins*, eds. Shafizadeh, F. *et al*. New York: Academic Press.

Brower, M.C., Tennis, M.W., Denzler, E.W. & Kaplan, M.M. (eds.) (1993). *Biomass Energy* (see Ch.3, Powering the Mid West – renewable electricity for the economy and the environment. Cambridge, MA: Union of Concerned. Scientists.

Brower, N. (1992). *Cool Energy – Renewable Solutions to Environmental Problems*. Cambridge, MA: MIT Press.

Brown, L.R. & Flavin, C. (1996). *State of the World – 1996*. New York: Norton for World Watch Institute. (a) Postel, S., Forging a sustainable water strategy, Ch. 3, pp. 40–59; (b) Abramovitz, J.N., Sustaining freshwater ecosystems, Ch. 4, pp. 60–77; (c) Gardner, G., Preserving agricultural resources Ch. 5, pp. 78–94; (d) Bright, C., Understanding the threat of bioinvasions, Ch. 6, pp. 95–113.

Brown, L.R., Flavin, C. & French, H. (1997). *State of the World – 1997*. New York: Norton for World Watch Institute. (a), Flavin, C., The legacy of Rio, Ch. 1, pp. 3–22; (b) Gardner, G., Preserving global cropland, Ch. 3, pp. 42–59; (c)

Abramovitz, J.N., Valuing Nature's services, Ch. 6, pp. 95–114; (d) Renner, M., Transforming security, Ch. 7, pp. 115–131.

Bull, S.R. (1991). The US Department of Energy Biofuels Research Program. *Energy Sources*, 13, 433–442.

Calvin, M. (1979). Petroleum plantations. In: *Solar Energy – Chemical Conversion & Storage*, eds. Hautala, R.R. *et al.*, pp. 1–30. Clifton, NJ: Humana Press.

Calvin, M. (1984). Renewable fuels for the future, a review, *Journal of Applied Biochemistry*, 6, 3–18.

Cassedy, E.S. (1993). Health risk valuations based on public consent. *IEEE Technology and Society Magazine*, 12 (Winter), 7–16.

Cassedy, E.S. & Grossman, P.Z. (1990). *Introduction to Energy – Resources, Technology and Society*, Chs. 3, 5, 12, Append. C. New York: Cambridge University Press.

Cassedy, E.S. & Grossman, P.Z. (1998). *Introduction to Energy – Resources, Technology, and Society*, 2nd edn. Cambridge: Cambridge University Press.

Cheremisinoff, P.N. & Morresi, A.C. (1976). *Energy from Solid Wastes.* New York: Marcel Dekker.

Curran, M.A. (1996). *Environmental Life Cycle Assessment.* New York: McGraw-Hill.

Demetrius, F.J. (1990). Brazil's National Alchohol Program – technology and development in an authoritative regime, New York: Praeger.

DOE (US Department of Energy) (1981). *Energy Technology and the Environment – Environmental Information Handbook.* Washington, DC: DOE. (a) Coal Gasification, Ch. 5, pp. 81–101; (b) Indirect Liquefaction of Coal, Ch. 6, pp. 103–147.

DOE (1991). *Programs in Utility Technologies, Solar, Thermal & Biomass Power Program Overviews, FY 1990–1991.* Washington, DC: DOE.

DOE (1993). *Biofuels – Program Plan FY 1992–1996.* DOE/CH10093–186, DE 93000036. Golden, CO: National Renewable Energy Laboratory.

DOE (1994a). Energy crops can provide large share of nation's fuel, *Biofuels Update*, 2, Issue 3. Washington, DC: DOE. (Report on USDOE Biofuels Technology.)

DOE (1994b). Switchgrass production costs estimated by ORNL. In: *Energy Crops Forum,* Fall 1994. Oakridge, TN: DOE Biofeedstocks Development Program, Oakridge National Laboratory. <http://www.esd.ornl.gov/bfdp/forum/94fall.htm#scosts>

DOE (1994c). *Biofuels: Project Summaries.* Report DOE/CH 10093–297, July. Golden, CO: National Renewable Energy Laboratory.

DOE (1995). *Task Force on Strategic Energy Research and Development, Annex 1: Technology Profiles*, Washington, DC: Secretary of Energy Advisory Board

DOE (1996a). (Report on US DOE Biofuels Technology.) *Biofuels Update*, 4, 1–4.

DOE (1996b). *Quarterly Bulletins.* (a) *AFDC Update, News of the Alternative Fuels Data Center.* Arlington, VA: Alternative Fuels Data Center; (b) *Biofuels Update. Report on NREL Biofuels Technology.* Golden, CO: National Renewable Energy Laboratory.

Dunn, P.D. (1986). *Renewable Energies: Sources, Conversion and Application.* London: Peregrinius Press.

Earl, D.E. (1974). *Forest Energy and Economic Development.* Oxford: Clarendon Press.

Ehrenshaft, A. (ed.) (1996). *Biofuels Feedstock Development Program (BFDP)*. Oakridge, TN: Oakridge National Laboratory. (a) Biofuels from trees: renewable energy research branches out, pp. 1–4, <http://www.esd.ornl.gov/bfdp/papers/misc/trees. htm>; (b) Energy crops and the environment, BFDP environmental research: soil and water conservation, water quality and wildlife diversity and habitat, pp. 1–3, <http://www.esd.ornl.gov/bfdp/papers/misc/cropenv.htm>; (c) Can biofuels improve our water quality?, energy crops forum Winter 1997 (newsletter) pp. 3–5, <http://www.esd.ornl.gov/bfdp//forum/97winter.htm>; (d) Biofuels from switchgrass: greener energy pastures, pp. 1–4, <http://www.esd.ornl.gov/bfdp/papers/misc/switgrs.html>; (e) Identifying the benefits of biomass plantings for wildlife, pp. 1–2, Energy Crops Forum Spring 1996, <http://www.esd.ornl.gov//bfdp/forum/96spring.htm>.

El Bassam, N. (1998). *Energy Plant Species.* London: James & James.

EPA (US Environmental Protection Agency) (1989). *National Air Quality and Emissions Trends*, EPA 450/491003. Research Triangle Park, NC: EPA

EPA (1992) *1992 EPA Project Summaries.* Washington, DC: EPA. (a) *Demonstration of Fuel Cells to Recover Energy from Landfill Gas: Phase 1 Final Report: Conceptional Study, EPA/600/SR92007*, February 1992: (b) *Landfill Gas Energy Utilization: Technology Options and Case Studies*, EPA/6600/SR92116, December 1992.

EUREC(EUREC Agency)(1996). *The Future for Renewable Energy Prospects and Directions*, London: James & James.

Evans, J. (1997). Worldwide experience with high yield forest plantations. *Biomass and Biomass Energy,* 13, 189–191.

Gardner, G. (1996). Asia is losing ground. *World Watch*, 9, Nov./Dec., 19–27.

Garg, D.P. (1987). Impact of biomass alternative energy strategies. In: *Progress in Solar Engineering,* ed. Goswami, D.Y., pp. 347–361. New York: Hemisphere Press (Harper & Rowe).

Goldemberg, J. Monaco, L.C. & Macedo, I.C. (1993). The Brazilian fuel alcohol program. In: *Renewable Energy – Sources for Fuels and Electricity*, Ch. 20, eds. Johansson, T.B., Kelly, H., Reddy, A.K.N. & Williams, R.H. Washington, DC: Island Press.

Gordon, D. (1991). *Steering a New Course – Transportation, Energy and the Environment.* Washington, DC: Union of Concerned. Scientists, Island Press.

Grassi, G. (1996). Potential impact of bioenergy on employment. *Biomass*, 1, 419–424.

Hall, D.O. (1994). Biomass energy options for Western Europe (OECD) to 2050, In: *Energy Technologies to Reduce CO$_2$ Emissions in Europe*, pp. 159–194. Paris: International Energy Agency, OECD.

Hall, D.O. & de Groot, P.J. (1986). Biomass for fuel and food, a parallel necessity, In: *Advances in Solar Energy*, Vol.3, ed. Boer, K.W. Boulder, CO: American Solar Energy Society.

Hall, D.O., Rosillo-Calle, F., Williams, R.H. & Woods, J. (1993). Biomass for energy: supply prospects. In: *Renewable Energy – Sources for Fuels and Electricity*, Ch. 14, eds. Johansson, T.B., Kelly, H., Reddy, A.K.N. & Williams, R.H., pp. 593–651. Washington, DC: Island Press.

Henderson, R.A. (1988). Net energy considerations, In: *Economic Analysis of Solar Thermal Energy Systems*, Ch. 7, eds. West, R.E. & Kreith, F. Cambridge, MA: MIT Press.

IEA (International Energy Agency) (1994). *Biofuels, Organization for Economic Cooperation & Development (OECD)*. Paris: International Energy Agency.

IGT (Institute of Gas Technology) (1978). *Symposium on Energy from Biomass and Wastes*, August, Chicago, IL. Papers including: (a) James, S.T. & Rhyme, C.W., Methane productions, recovery and utilization from landfills, pp. 317–324; (b) Stearns, R.P. *et al.*, Recovery and utilization of methane gas from a sanitary landfill, pp. 325–344; (c) Rice, F.C., Commercial production of pipeline quality gas at the Palos Verdes Landfill, pp. 345–352; (d) Madewell, J.F., Quads from biomass, pp. 837–845; (e) Mariani, E.O. The eucalyptus energy farms as a renewable source of fuel, pp. 29–38.

Kaltschmitt, M. Reinhardt, G.A. & Stelzer, T. (1996). LCA of biofuels under different environmental aspects, In: *Biomass for Energy and Environment*, Vol. 1, eds. Chartier, P., Ferrero, G.L., Henius, U.M., Hultberg, S. & Sachau, J., pp. 369–386. London: Pergamon, Elsevier.

Keeney, D.R. (1996). National food production concerns: stable food and agriculture. *Renewable Resources Journal*, Spring, 7–10.

Kishor, N. & Constantino, L. (1994). Sustainable forestry: can it compete? *Finance and Development*, Nov., 36–39.

Kohl, W.L. (ed.) (1990). *Methanol as an alternative fuel choice: an assessment*. Washington, DC: Johns Hopkins Foreign Policy Institute.

Korbitz, W. (1998). From the field to the fast lane. *Renewable Energy World*, 1, 32–37.

Kort, J., Collins, M. & Ditsch, D. (1998). A review of soil erosion potential associated with biomass crops. *Biomass and Biomass Energy*, 14, 4, 351–359.

Krupnick, A.J. & Portney, P.R. (1991). Controlling urban air pollution: a benefit–cost assessment. *Science*, 252, 522–528.

Lehmann, H., Payke, R., Pfluger, A. & Reetz, T. (1996). Sustainable land use in the European Union, In: *Biomass for Energy and Environment*, Vol. 3, eds. Chartier, P., Ferrero, G.L., Henius, U.M., Hultberg, S. & Sachau, J., pp. 1727–1732. London: Pergamon, Elsevier.

Lettinga, G. & van Haandel, A.C. (1993). Anaerobic digestion for energy production and environmental protection. In: *Renewable Energy – Sources for Fuels and Electricity*, Ch. 19, eds. Johansson, T.B., Kelly, H., Reddy, A.K.N. & Williams, R.H. Washington, DC: Island Press.

Lynd, L.R. (1989). Production of ethanol from lignocellulosic materials using thermophilic baceria. In: *Advances in Biochemical Engineering and Biotechnology*, Vol. 38, ed. Fiechter, pp. 2–52. A. Berlin: Springer-Verlag.

Lynd, L.R., Cushman, J.H., Nichols, R.J. & Wyman, C.E. (1991). Fuel ethanol from cellulosic biomass. *Science*, 251, 1318–1323.

Magee, R.J. & Kosaric, N. (1985). Bioconversion of hemicellulosics. In *Advances in Biochemical Engineering and Biotechnology*, Vol. 32, ed. Fiechter, A. pp. 61–93. Berlin: Springer-Verlag.

Maser, C. (1994). *Sustainable Forestry, Philosophy, Science and Economics*. Delray Beach, FL: St Lucie Press.

Mattoon, A.T. (1998). Paper forests. *World Watch*, March/April, 20–28.

Meynell, P.J. (1982). *Methane: Planning a Digester*, 2nd edn. Dorchester, NY: Prism Press.

Miles, T.R. & Miles, T.R. Jr (1989). Overview of biomass gasification in the USA. *Biomass*, 18, 163–168.

Moreno, R. & Bailey, D.G.F. (1989). *Alternative Transport Fuels from Natural Gas.* Washington, DC: The World Bank.

Mukunda, H.S., Dasappa, S. & Shrinivasa, U. (1993). Open-top wood gasifiers. In: *Renewable Energy Sources for Fuels and Electricity*, Ch. 16, eds. Johansson, T.B., Kelly, H., Reddy, A.K.N. & Williams, R.H. Washington, DC: Island Press.

NRC (National Research Council) (1990). *Fuels to Drive Our Future.* Washington, DC: National Academy Press.

NREL (National Renewable Energy Laboratory) (1992). *Conservation and Renewable Energy Technologies for Transportation.* Washington, DC: DOE.

NREL (1993). *Profiles in Renewable Energy – Case Studies of Successful Utility Sector Projects.* DOE/CH10093–206, DE3000081, Golden, CO: National Renewable Energy Laboratory.

NREL (1995). (Report on USDOE Biofuels Technology.) *Biofuels Update*, Newsletter, 3, Issue 1.

Office of Technology Assessment (1990). *Replacing Gasoline – Alternative Fuels for Light-Duty Vehicles.* Report DTA–365. Washington, DC: US Congress.

Ogden, J.M. & Nitsch, J. (1993). Solar hydrogen. In: *Renewable Energy – Sources for Fuels and Electricity*, Ch. 22, eds. Johansson, T.B., Kelly, H., Reddy, A.K.N. & Williams, R.H., pp. 925–1009. Washington, DC: Island Press.

Overend, R.P. (1998). Biomass gasification: a growing business. *Renewable Energy World*, 1, 27–31.

Pagiola, S., Kellenberg, L., Vidaeus, L. & Shivastava, J. (1998). Mainstreaming biodiversity in agricultural development. *Finance & Development*, March, 38–41.

Patterson, W. (1994). *Power from Plants – The Global Implications of New Technologies for Electricity from Biomass.* London: The Royal Institute of International Affairs, Earthscan Publications.

Penner, S.S., Wiessenhalm, D.F. & Li, C.P. (1987). Mass burning of municipal wastes. *Annual Review of Energy*, 12, 415–444.

Perrings, C., Grubb, M., Walker, J., Ruxton, R., Glenny, T. & Herring, H. (eds.) (1995). *Biodiversity Loss: Economic and Ecological Issues*, Cambridge: Cambridge University Press.

Pimental, D. & Giampietro, M. (1994). *Food, Land, Population and the US Economy, Carrying Capacity.* Washington, DC: Network <http://dieoff.org>, pp. 1–31.

Ranney, J.W. (1986). New technologies in the production of woody crops for energy in the United. States. In: *Advances in Solar Energy*, Vol. 3, ed. Boer, K.W. Boulder, CO: American Solar Energy Society.

Ranney, J.W. & Cushman, J.H. (1991). Energy from biomass. In: *The Energy Sourcebook – A Guide to Technology, Resources and Policy*, pp. 299–311, New York: American Institute of Physics.

Reed, T.B. (1993). Principles of biomass gasification. In: *Advances in Solar Energy*, Vol. 2, Ch. 3, eds. Boer, K.W. & Duffie, J.A., pp. 125–174. Boulder, CO: American Solar Energy Society.

Rissler, J. & Mellon, M. (1996). *The Ecological Risks of Engineered Crops.* Cambridge, MA: MIT Press.

Sage, R.B. (1998). Short rotation coppice for energy: toward ecological guidelines. *Biomass & Biomass Energy*, 15, 1, 39–47.

Saha, B.C. & Woodward, J. (eds.) (1997). Fuels and chemicals from biomass. *ACS Symposium Series: Advanced Bioethanol Production Technologies – A Perspective*. Washington, DC: American Chemical Society.

Schell, D.J., McMillan, J.D., Philippidis, G.P., Hinman, N.D. & Riley, C. (1992). Ethanol from lignocellulosic biomass, In: *Advances in Solar Energy*, Vol. 7, Ch. 10, ed. Boer, K.W., pp. 73–448. Boulder, CO: American Solar Energy Society.

Schippler, L. & Sperling, D. (1994). Ethanol's cost far outweighs the benefit (Letter to the Editor). *The New York Times*, April 19.

Smith, W.H. & Frank, J. (eds.) (1988). *Methane from Biomass – A System Approach*. New York: Elsevier.

Speece, R.E. (1985). Environmental requirements for anaerobic digestion of biomass. In: *Advances in Solar Energy*, Vol. 2, Ch. 2, ed. Boer, K.W. Boulder, CO: American Solar Energy Society.

Stobough, R. & Yergin, D. (1979). Energy future. In: *Solar America*, Ch. 7. New York: Random House.

Strauss, C.H. & Grado, S.C. (1991). Energy and financial costs for SRIC woody biomass. In: *Solar World Congress*, ed. Adler, M.E. *et al.*, pp. 837–942. Oxford: Pergamon Press.

Tuscan, G.A. (1998). Short rotation woody crop systems in the United. States: what do we know and what do we need to know. *Biomass and Biomass Energy*, 14, 4, 307–315.

USDA (US Department of Agriculture) (1989). *Ethanol's Role in Clean Air*. Washington, DC: USDA Backgrounder Series.

USDA (1992). *Proceedings of a Meeting on Technology for Expanding the Biofuels Industry*, April, Chicago. Sponsored by the USDA, the US Department of Energy and the Renewable Fuels Association.

USDA (1993). Emerging technologies in ethanol production. *Agriculture Information Bulletin No. 663*. Washington, DC: Economic Research Service.

Vallander, L. & Eriksson, K.-E.L. (1990). Production of ethanol from lignocellulosic materials: state of the art. In: *Advances in Biochemical Engineering/Biotechnology*, Vol. 42, *Bioprocesses & Applied. Enzymology*, ed. Fiechter, A., pp. 63–95. Berlin: Springer-Verlag

van der Bijl, G. (1996). Sustainability of production and use of biomass for European energy supply. In: *Biomass for Energy and Environment*, Vol. 1, eds. Chartier, P., Ferrero, G.L., Henius, U.M. Hultberg, S.& Sachau, J., pp. 387–392. London: Pergamon, Elsevier.

Warmer (World Action for Recycling Materials & Energy from Rubbish) Campaign (1991–94). Various articles in *Warmer Bulletin*. Kent, UK: Warmer Campaign. (a) Salzburg chooses anaerobic digestion (1993). *Warmer Bulletin*, 37 (May), 6; (b) Combined. heat and power the solution? (1992). *Warmer Bulletin*, 32 (Feb.), whole issue; (c) Technical: energy from wastes (1992). *Warmer Bulletin*, 34 (Aug.), 18–19; (d) Green power scheme – Leicestershire County, Base Load Systems Landfill Gas Scheme (1991). *Warmer Bulletin*, 29 (May), 20; (e) Mistry, P. Anaerobic digestion of municipal solid waste (1991). *Warmer Bulletin*, 29 (May), 10; (f) Prevention rather than cure: gasification of refuse derived fuel (1991). *Warmer Bulletin*, 29 (May), 18; (g) Anaerobic digestion (Warmer Information Sheet) (1994). *Warmer Bulletin*, 40 (Feb.); whole issue.

Wilen, C. & Rautalin, A. (1996). Safe handling of biomass fuels in IGCC power production. In: *Biomass for Energy and Environment*, Vol. 1, ed. Chartier, P. *et al.,* pp. 170–175. London: Pergamon, Elsevier.

Williams, R.H. & Larson, E.D. (1993). Advanced gasification–based biomass power generation. In: *Renewable Energy – Sources for Fuels and Electricity*, Ch. 17, eds. Johansson, T.B., Kelly, H., Reddy, A.K.N. & Williams, R.H. Washington, DC: Island Press.

Williams, S. & Bateman B.G. (1995). *Power Plays – Profiles of America's Independent Renewable Electricity Developers.* Washington, DC: Investor Responsibility Research Center.

WRI (World Resources Institute) (1996*). World Resources – A Guide to the Global Environment (1996–97),* Ch. 9 (Forests and Land Cover), 10 (Food and Agriculture). New York: WRI/UNEP/UNDP/World Bank, Oxford University Press,

Wyman, C.E., Bain, R.L., Hinman, N.D. & Stevens, D.J. (1993). Ethanol and methanol from cellulosic biomass. In: *Renewable Energy – Sources for Fuels and Electricity*, Ch. 21, eds. Johansson, T.B., Kelly, H., Reddy, A.K.N. & Williams, R.H. Washington, DC: Island Press.

Further reading

Anon. (1989). The greenhouse effect: a tropical forestry response (Short communication). *Biomass*, 18, 73–78.

ASES (American Solar Energy Society) (1992). *Alternative Transportation Fuels and Vehicles.* Transcript of Roundtable held April 28, Washington, DC: American Solar Energy Society.

Bradshaw T. (1996). Poplar Molecular Genetics Cooperative, In: *Energy Crops Forum Summer 1996.* <http://www.esd.ornl.gov/bfdp/forum/96sum.html> Oakridge, TN: DOE Biofuels Feedstock Development Program, Oakridge National Laboratory.

Dayal, M. (1989). *Renewable Energy: Environment and Development.* Delhi, India: Konark.

Groombridge, B. (ed.) (1992). *Global Biodiversity – Status of the Earth's Living Resources.* London: World Conservation Monitoring Centre, Chapman & Hall.

Johansson, T.B., Kelly, H., Reddy, A.K.N. & Williams, R.H. (1993). *Renewable Energy – Sources for Fuels and Electricity.* Washington, DC: Island Press.

OECD (Organization for Economic Cooperation & Development) (1995). *Agricultural Policy Reform and Adjustment for Japan.* Paris: Organisation for Economic Cooperation & Development.

Overend, R.P. (1993). Biofuels for a cleaner future. *Solar Today*, Jan./Feb., 11–14.

Shiva, V. (1993). *Monocultures of the Mind: Perspectives on Biodiversity and Biotechnology.* Penang: Third World Network.

Sweeney, J.L. (1990). *Projected Costs of Alternate Liquid Transportation Fuels.* Stanford, CA: Stanford University Center for Economic Policy Research.

Tenner, E. (1996). *Why Things Bite Back – Technology and the Revenge of Unintended Consequences,* Ch. 7, Acclimatizing pests: vegetables. New York: Alfred. A. Knopf.

Wood, K. (1991). Wood as a home heating fuel. *Solar Today*, Sept./Oct., 17–19.

3 Windpower

3.1 Background

Modern wind power is already supplying electricity to utility grids and remote-site users around the world. It is a maturing technology, with its ultimate limitations only a consequence of its intermittency. It is presently cost competitive only at "good" wind sites. Energy production costs are already in the competitive range at those good wind sites. "Wind farms" at those sites appear to have a bright future for becoming part of the electric generation mix of many utilities, especially in North America and Northern Europe.

Wind energy has been used throughout recorded history for sailing, pumping water, and grinding grains (Kealy, 1987; Hills, 1994). Windmills, of one sort or another, are known to have provided pumping and milling energy in the earliest Mediterranean and Eastern civilizations and in medieval Europe. With the advent of fossil-fired energy in 18th century Europe and electricity throughout the industrial countries in the late 19th century, wind and water power fell more and more into disuse, persisting only in the lowlands of northern Europe into the 20th century (Sesto & Casale, 1994). While the multibladed windmill was still a familiar sight in rural areas of the American Southwest, South Africa, and Australia, wind power had become virtually extinct by the middle of the 20th century in most of the industrial countries (Dodge & Thresher, 1989). Modern windpower, like modern hydropower, delivers electricity rather than mechanical power. The modern wind turbine consists of a set of "impellers", much like aircraft propellers, which drive the shaft of an electric generator. Figure 3.1a shows the cross-section of a modern turbine, together with all auxiliary equipment, including controls, sensors, shock absorbers, and the like. Several of the functions of these other components will become evident in the text to follow.

While some experimentation in the USA with wind-driven electric generation took place in the 1940s, the beginning of the current efforts for wind turbines was in the energy-crisis period of the 1970s (De Meo & Steitz, 1990). Taking advantage of advances in modern aerodynamics, a prototype wind turbine series, aiming for multimegawatt outputs, was built and tested in the USA under government (DOE and NASA) sponsorship. This series progressed to MOD-2, which had a 300 ft (91 m) diameter, two-blade rotor (Fig. 3.1b) and a 2.0–2.5 MW output capacity; this turbine first operated in 1981. Several other MOD-2 units were operated on utility systems in the early 1980s, but funding was soon cut and the federal

(a)

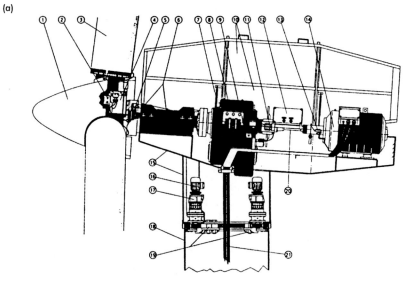

1. Nose cone	8. Coaxial gearbox
2. Hub	9. Hydraulics
3. Blades	10. Nacelle
4. Hydraulics	11. Brake
5. Slew ring system	12. Controls
6. Main shaft	13. Vibration sensor
7. Shock absorber	14. Generator

15. Bed plate
16. Yaw motor
17. Yaw gear
18. Tower
19. Yaw system
20. Transmission shaft
21. Power cables

(b)

Figure 3.1
Modern wind turbine.
(a) Cross-sectional
drawing. Reprinted
with permission from
Johansson, T.B. *et al.*,
(eds.) (1993).
*Renewable Energy –
Sources for Fuels and
Electricity*, Island Press,
Washington, DC and
Covello, CA.) (b) Large
wind turbine, NASA
MOD-2 (2.5 MW
capacity). (Courtesy of
the American Solar
Energy Association.)

demonstration program died, having spent $290M over the period 1974–81 (Frankel, 1986).

In place of the funded R&D projects came a government-promoted, cost-sharing arrangement with utilities and turbine manufacturers for demonstration of larger designs in the MOD series. The most notable outcome of this later program was the installation and operation of MOD-5B, with a 320 ft (97 m) diameter rotor and a 3.2 MW output for the Hawaiian Electric Co. While some progress was made in design and valuable operating experience was gained in these demonstrations, they fell victim to the same loss of interest that blighted the solar projects in the 1980s. And like solar, wind technologies did not reach economic competitiveness in the prevailing energy markets.

The experience of the US MOD series, nonetheless, led to technical progress that was incorporated into successive generations of machines (Hock *et al.*, 1992). The early wind turbines were subject to mechanical fatigue failures (Dodge & Thresher, 1989; De Meo & Steitz, 1990; Gipe, 1995b) and could operate only over a very limited range of wind speeds. A principal category of failure involved the blades, which are subject to sudden twist and flexing stresses in ever recurring wind gusts, leading to fatigue failure. Subsequent blade designs have utilized lightweight, high-strength materials, as used in sail-boat masts or light aircraft. In addition, the design trend for later wind-farm applications was toward smaller wind turbines in larger numbers (Figs. 3.2a,b). These machines, with rotor diameters around 30–100 ft (10–30 m), have suffered far less stress and fatigue failures. The relationship between rotor diameter and output power, in this range of machine sizes, is shown in Fig. 3.3.

Some of the earlier MOD series machines were designed to operate "downwind," with the rotor downwind of the Nacelle (generator housing). Downwind machines have the advantage of being self-orienting with the wind direction. Their disadvantage, however, is that the rotors are affected badly by wind turbulence around the tower structure. Subsequent designs were "upwind", where orientation into the wind was achieved by tail fins (as for historic windmills) or by modern feedback control.

3.2 Operating principles

Modern wind turbines are designed on the aerodynamic principle of lift (Eldridge, 1980; Dunn, 1986) similar to that of aircraft wings or sail boats (tacking upwind). Lift forces are created on an aerofoil (a tapered, tear-drop in cross-section) when its leading edge is oriented at a small angle to the direction of the incoming wind. Lift is a consequence of the differential in air pressure between the two sides resulting from the differing path lengths the air must take in flowing around the foil. If the angle of attack of the foil is too large, the wind simply pushes against the windward surface, exerting a drag force in the direction of the wind but no lift of the foil. When the drag is too great relative to the lift, the condition of a stall occurs, as can happen with aircraft wings.

(a)

Figure 3.2
Medium-sized wind turbines. (a) WindMaster turbines at Altamont Pass, CA. (b) Vestas turbines at Techachpi, CA. (Courtesy of the American Wind Energy Association.)

The lift force (only) turns windmill blades for useful power output. The angle of attack will vary if the aerofoil is moving, as happens with wings and sails under motion. This effect may be viewed as creating an "apparent wind," made up of velocity components of the original wind together with the motion of the aerofoil itself. For windmill blades, the rotational motion creates an apparent wind that is shifted the maximum amount at the blade tips and not at all at the center of rotation. Therefore, an

(b)

important parameter for windmill-blade operation is the ratio of tip speed to wind speed.

The pitch (angle of attack) of the windmill blade can be designed so that the entire rotor delivers maximum power in a given range of tip–speed ratios (Dodge & Thresher, 1980). This can be done either by making the pitch adjustable in operation or by designing a permanent twist in the blade over its length (Hock *et al.*, 1992). A given fixed-blade design has a maximum of output power and torque only in a particular range of rotational speeds. Generally, a tip–speed ratio of 10:1 or more, is sought in order to achieve higher rotational speeds. The rotational speed can be controlled in operation with variable-pitch blades (although not with high precision) using these speed characteristics plus other features such as aerodynamic spoilers or mechanical brakes (Smith, 1987). It would take other means, as we will presently see, to meet the frequency-consistency requirements adequately of the a.c. power systems that the wind turbines are to feed.

The same aerodynamic principles apply to wind machines with vertical axes (Fig. 3.4). The best-known vertical-axis machine is the Darrieus, which was invented in 1927 (Hickok,1975; Dodge & Thresher, 1989). In these machines, horizontal lift forces are developed on the vertical sections of the bowed foils, which turn the entire structure. (The bowed shape results in part from centrifugal forces, under rotation, of the flexible blades.) While these machines are relatively efficient, they have a disadvantage of not being self starting.

3.3 Operating experience

A number of operational problems became evident in the US MOD

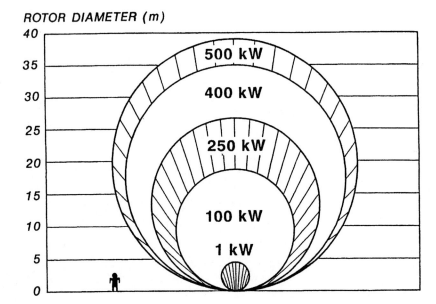

Figure 3.3
Rotor diameter versus
output power capacity.
(*Source*: Boer (1994),
American Solar Energy
Association.)

demonstrations that continued into the early 1980s. In addition to the
blade failures already mentioned, fatigue failures occurred in the turbine
itself and in the support structures from the yaw and teetering motions that
occur incessantly and randomly in variable and turbulent winds. The
(transported) energy in the wind rises rapidly (as the third power) with
wind speed. Wind machines are designed to operate with velocities up to
about 35 mph (16m/s), but must be able to survive gales (in "stowed" or
feathered positions) of 100 mph (45m/s). The design responses to these
mechanical failures were trends to lighter rotors, gimbaled mountings, and
more flexible supporting structures in the American designs (Hock, *et al.*,
1992; NERL, 1992), but to greater strength and rigidity in some subsequent European designs. The ultimate test of all these designs was to be
able to erect wind machines that would withstand *all* wind conditions, and
to do so at a competitive cost.

It became apparent to some designers that the multimegawatt output
range, motivated out of the conventional wisdom of economy of scale, was
ill advised (De Meo & Steitz, 1990; Grubb, 1993). Not only were the unit
capital costs too high (up to $2000/kW$_e$), but high outage rates resulted in
capacity factors too low to pay off the investments with the required
returns. Gipe (1995) called this "the death knell of the giants". The next
generation of wind machines changed all this by moving to smaller machines; leading designs were produced by the Danish manufacturers in a
submegawatt "intermediate" range of capacities (50–500 kW) (Hock *et al.*,
1992; Brower *et al.*, 1993; Wager, 1994).

This next generation started in the early 1980s (De Meo & Steitz, 1990;
Lamarre, 1992), driven financially by government incentives rather than as
a totally sponsored federal demonstration effort as before. Incentives for
wind and other forms of renewable generation were also introduced in

Figure 3.4
Darrieus vertical-axis
wind machine at
Altamont Pass, CA,
manufactured by the
Flowind company.
(Courtesy of the
American Wind Energy
Association.)

Europe in the 1980s (Evans, 1992; Gipe, 1995a,b), stimulating development efforts there. Some government-sponsored, wind-turbine design and
development in the USA was carried over from the MOD series, but the
new designs were built and put into operation by private American manufacturers, such as US Wind Power, Inc., and a few European firms,
motivated financially by the prospects of tax investment credits and
PURPA operating guarantees in the USA and the market prospects for the
farming community in Denmark.

Such activity in the USA was focused in California, where a 50% state
energy-tax credit could be added to federal credits (15% energy credit and

10% investment) creating a particularly advantageous tax incentive package. In addition, the State of California would guarantee prices for renewable energy for several years after the start of a project (Lamarre, 1992). In Denmark, wind generators were allowed to avoid the carbon tax on fossil-fuel plants and farmers using wind power were excused from the value-added tax (VAT). By the time that federal tax credits in the USA had ended in 1985, a new generation of wind turbines had been established on the "wind farms" of California.

Even this generation of smaller wind machines had its growing pains: machine failures and low-capacity factors occurred with earlier American models (Smith, 1987; De Meo & Steitz, 1990). During the 1980s, European makes, such as the Danish-made Vesta and Nordtank machines, which were typically three-bladed, heavier, and more rigid in design than the American models, made inroads into the domestic market in the USA. Later American models of three-blade designs (Fig. 3.2a), coming after a shakeout of the wind-machine manufacturers, have achieved impressive operating records on the California wind farms and in pilot clusters in northern Europe, but not before Danish machines had captured 50% of the world market (Sesto & Casale, 1994; Gipe, 1995a,b). Operating availabilities over 97% and annual capacity factors approaching 40% have been attained using these intermediate-capacity machines. They are operated on these farms utilizing centralized monitoring and control of whole arrays of machines, together with centrally managed maintenance. Operation and maintenance costs have been lowered to the neighborhood of 1¢/kW-hr.

The investment costs for both US and European wind turbines fell during the late 1980s (De Meo & Steitz, 1990; DOE, 1992; Heier *et al.*, 1994), approaching a unit capacity cost of $1000/kW by the early 1990s, which is comparable to capacity costs of conventional electric plants. Further incentives for investments came in the 1992 federal energy bill (the *National Energy Policy Act* of 1992), which allowed $0.015/kW-hr production incentives to utilities for renewable energy, although the 105th Congress failed to renew this incentive in 1998. The overall objective with federal (DOE) design sponsorship (NREL, 1992) and project promotion by EPRI (Electric Power Research Institute) has been a competitive production cost of $0.05/kW-hr (at the busbar), but this goal was revised downward to at least $0.035 in 1997 (DOE, 1997). In 1988, the cost of wind generation was $0.10/kW-hr (DOE, 1992, 1994) and had fallen below $0.09/kW-hr by 1994 (Sesto & Casale, 1991; Gipe, 1995a,b), a competitive cost of electricity at the generator busbar in many parts of the USA. In 1997, a 20-year contract was signed with Enron Corp. to supply 45 MW (average power) to an upper Midwest utility at rates below $0.05/kW-hr (Anon., 1997).

The California wind farms – operating now for more than a decade – have demonstrated the feasibility and advantages of the new approach of using many medium-sized machines rather than a few large ones (Smith, 1987). A typical machine in this category is the US Windpower model 33M-VS, which has an output power rating of 400 kW and a rotor diameter of 108 ft (33 m), which is similar in size and output to the

Danish-made model Vestas-DWT (Sesto & Casale, 1994). Other US manufacturers include Zond and Flo Wind (Williams & Bateman, 1995). The largest American-made machine was 750 kW capacity, made by Zond (now owned by Enron, Corp.) while European makes are following a trend even to over 1 MW (EUREC, 1996; Parsons, 1998), leading to structural and economic limits. Operational advantages have been found with smoother aggregate electric outputs from large clusters of such machines, in addition to the higher reliability of the aggregate of machines (Beyer *et al.*, 1993).

3.4 Wind resources

The success of the Altamont Pass wind farms emphasizes the importance of siting for the wind resource. It is well known that the sources of winds on a global basis are the differential heating of the equatorial and polar regions, plus the earth's rotation (Eldridge, 1980; Cavallo *et al.*, 1993). These forces create the prevailing winds, such as the trade winds in the tropical latitudes and the prevailing westerlies in the temperate latitudes. These prevailing winds are modified on a regional basis by the presence of land masses, oceans, and continental topographies. The consistent winds of the Altamont Pass, for instance, result from both the prevailing westerlies and the "sea breeze" caused by differential solar heating of the land versus the water near a coastline. These winds are enhanced by being channeled through the Pass in the hills east of San Francisco Bay.

Regional wind resources, resulting from these widespread forces, have been mapped on broad scales (Eldridge, 1980) in terms of wind "availability", expected wind speeds, or average wind-power densities on regional scales. Figure 3.5a shows worldwide wind availability (referring to wind-resource availability not turbine operational readiness); here, *availability* measures the annual energy (in kW-hr) obtainable per installed capacity (kW) of wind machine. Availability may also be interpreted as the annual hours of operation at full (kW) capacity, where it might be recalled that there are (24 x 365) 8760 hours per year. Therefore, it is only in the 5000 kW-hr/kW regions that equivalent full-capacity operation is expected for 5000 hours or more annually, which would, in turn, yield competitive capacity factors for investment returns. We see on a world scale the highest wind availability in coastal regions, where the land–sea effect operates (see above), and in mountainous regions, where the stronger winds at higher altitudes can be utilized. Good wind areas, however, are also found in mid-continent regions and at lower elevations.

Figure 3.5b shows the geographic distribution of expected wind speeds across the continental USA, showing the regional influences on wind availability. The resources along coastlines are very well illustrated in the case of northern Europe (Fig. 3.5c), the home of early windmills and the sites of modern wind-turbine demonstrations. Expected wind velocities of 15 mph (7m/s) or more are rated as excellent for wind turbines 10 m above ground level; whereas those less than 10 mph (5m/s) are considered poor.

(a)

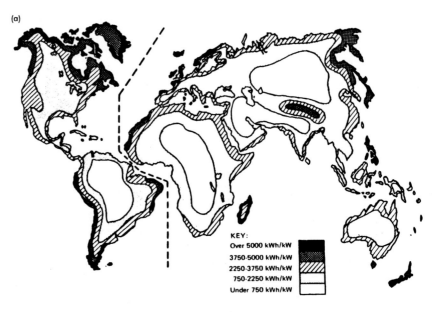

KEY:
Over 5000 kWh/kW
3750-5000 kWh/kW
2250-3750 kWh/kW
750-2250 kWh/kW
Under 750 kWh/kW

Figure 3.5
Wind resources. (a)
Worldwide wind
availability.
(Courtesy of the
National Science
Foundation, 1975.) (b)
Annual average wind
power in the USA. (The
classes correspond to
those in the
table.)(*Source*: Pacific
Northwest Laboratory,
1986.) (c) annual
average wind power in
Europe. (*Source*: Risoe
National Laboratory of
Denmark.) (d) Wind
resources in the upper
Midwest (increasing
class number indicates
(increasing wind
power; the classes are
detailed in the table
with (a).) (e) Wind map
for the upper Midwest
revised from (d) to
include local effects
such as terrain
elevation and
roughness. Parts (d)
and (e) reprinted with
permission from
Brower, M.C. *et al.*,
*Powering the Midwest
– Renewable Electricity
for the Economy and
the Environment*, Union
of Concerned
Scientists, Cambridge,
MA.)

(b)

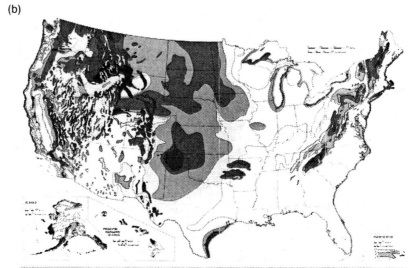

Wind resources at two heights above ground level

Wind power class	10 m (33 ft)			50 m (164 ft)		
	Wind power W/m²	Speed m/s	(mph)	Wind power W/m²	Speed m/s	(mph)
1	0	0	(0)	0	0	(0)
2	100	4.4	(9.8)	200	5.6	(12.5)
3	150	5.1	(11.5)	300	6.4	(14.3)
4	200	5.6	(12.5)	400	7.0	(15.7)
5	250	6.0	(13.4)	500	7.5	(16.8)
6	300	6.4	(14.3)	600	8.0	(17.9)
7	400	7.0	(15.7)	800	8.8	(19.7)
	1000	9.4	(21.1)	2000	11.9	(26.6)

 Ridge crest estimates (local relief) 1000 ft

(c)

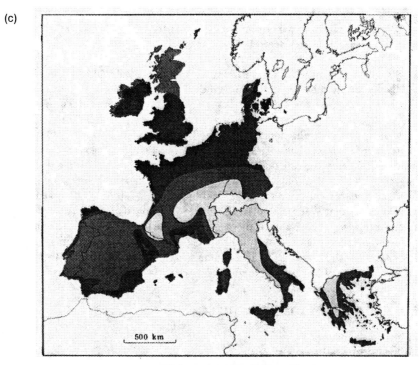

Wind resources¹ at 50 m above ground level

	Sheltered terrain		Open plain		At a sea coast		Open sea		Hills and ridges	
	m/s	W/m²	m/s	W/m²	m/s	W/m²	m/s	W/m²	m/s	W/m²
	>6.0	>250	>7.5	>500	>8.5	>700	>9.0	>800	>11.5	>1800
	5.0–6.0	150–250	6.5–7.5	300–500	7.0–8.5	400–700	8.0–9.0	600–800	10.0–11.5	1200–1800
	4.4–5.0	100–150	5.5–6.5	200–300	6.0–7.0	250–400	7.0–8.0	400–600	8.5–10.0	700–1200
	3.5–4.5	50–100	4.5–5.5	100–200	5.0–6.0	150–250	5.5–7.0	200–400	7.0–8.5	400–700
	<3.5	<50	<4.5	<100	<5.0	<150	<5.5	<200	<7.0	<400

(d)

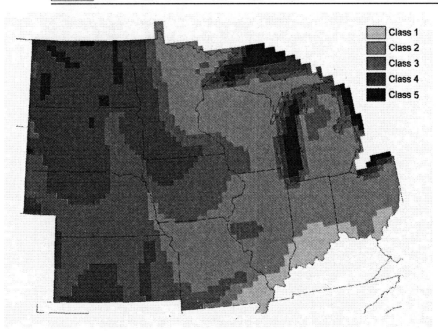

Class 1
Class 2
Class 3
Class 4
Class 5

(e)

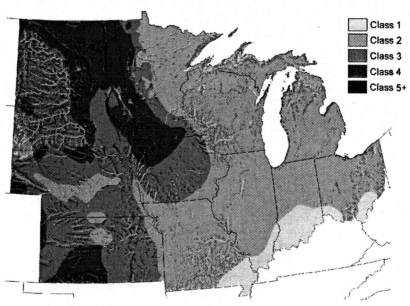

Class 1
Class 2
Class 3
Class 4
Class 5+

Figure 3.5 *cont.*

The units for this map are given also in terms of annual average power densities, which can be interpreted as the potential wind-power density (e.g. W/m^2) to deliver power to the area (m^2) swept by the turbine rotor blades. (Note that the relationship of wind power with wind velocity is not linearly proportional or even quadratic but varies rather as the third power of velocity.) The various regions are also categorized in terms of wind classes, accounting for wind power or speed at higher elevations above the ground (see the table with Fig. 3.5b).

We can see, as with wind availability, the highest average wind speeds occurring at coastlines and in mountainous regions. Actually, the use of average wind speeds as a measure of regional prospects for wind generation may not be as appropriate as the availability measure because it would overemphasize the high wind speeds (35 mph, or 16m/s, and above) when the turbines cannot be operated. In addition, it would not account for the constancy (or statistical "duration") of the wind power, which is indicated by the availability measure. Indeed, some of the US sites chosen for utility wind projects, after California, have been in the Upper Midwest (Lamarre, 1992; Swisher & Gipe, 1992; UWIG, 1992; Brower, *et al.*, 1993; Wager, 1994), where the wind velocities are moderately high, but not very high. The northwest coasts of Europe also have been identified for further wind development both on and off shore (Evans, 1992; Grubb & Meyer, 1993; EUREC, 1996).

The US Upper Midwest is the focus of a revealing regional study of wind and other renewable resources (Brower *et al.*, 1993; Wager, 1994). In assessing wind resources in the Great Plains region, these authors employed a computerized mapping procedure that shows the average wind-power density in a geographical grid distribution for blocks of land approximately about 40 km × 40 km (25 mile × 25 mile) dimensions (Fig. 3.5d). Each block on the map is designated by a *wind class*, according to the average power densities found in that subregion at specific heights above the ground (see the table with Fig. 3.5b).

The additional specification of height for wind class is important, since wind speeds generally increase with height, and the higher the mounting hub of the wind machine the greater will be the collection of energy. Accounting for such factors, each block on the map is designated by a wind class, thereby indicating the suitability for wind generation there. Wind classes of 5 and higher are considered productive for generation using the existing wind turbine technology and classes 3 and 4 promising after expected improvements in the technology.

The pattern of power densities seen in Fig. 3.5d for the Midwest results from large-scale air movements at mid-continent, prevailing from west to east, with again the highest wind speeds occurring at high altitudes and over expanses of water. What is not indicated, however, are the local effects caused by (moderate) changes in terrain elevation and terrain roughness (from trees and other vegetation). Such effects are included in the revised map shown in Fig. 3.5e, which is also mapped on a much finer scale (1 km x 1 km) and which excludes urban areas and the open waters of the Great Lakes as inappropriate for wind farms.

What we see on the revised map in Fig. 3.5e, are those regions of the Upper Midwest that are technically and socially feasible for wind farms and we can distinguish which are the most promising for development. The more promising regions (darker shaded on the map) generally are the flat or gently rolling prairie croplands of Iowa, Kansas, and the Dakotas, with large areas in wind class 4 and many spotted class 5 areas. Other studies show Montana and Wyoming as also very promising (Grubb & Meyer, 1993; Williams & Bateman, 1995). In most of these rural areas, dual use of wind farms with agriculture – so successful in California – appears possible. With such dual use, land owners are able to supplement their income with land rents for every acre occupied by wind generators (Wager, 1994). The collective potentials for windpower generation for each of the Midwest states is shown in Table 3.1. This indicates that those 12 states are estimated to have the resource potential to supply over 11 times the electric (energy) demand from wind generation, even when environmental and urban land exclusions are taken into account (Elliott & Schwartz, 1993). (However, only with mass electricity storage technology could this be used to supply a high percentage of that electric energy demand; see Chapter 5.)

Considerations of regional siting should not override the importance of local siting, however, in deciding the placement of wind turbine or farms. Local, so-called micro siting, factors include topography in the proximity (e.g. nearby hills and passes), proximity (a few kilometers) to large bodies of water, local height (above the surrounding terrain), and "fetch" (kilometer-scale distances of unobstructed spaces to windward). In general, local wind patterns are difficult to predict, since air movements are random, turbulent phenomena and must be studied carefully at any particular microsite. Performance gains from increased wind speeds and lowered turbulence can be achieved at most sites using higher towers (Cavallo, *et al.*, 1993), an approach under consideration for the next

Table 3.1. *Wind energy potentials for the upper Midwest of the USA (with environmental and urban exclusions)*

State	Class 3			Class 4–6		
	Thousand megawatts peak	Billion kW-h annual	Multiple of electric demand	Thousand megawatts peak	Billion kW-h annual	Multiple of electric demand
Illinois	26	53	0.40	0	0	0.00
Indiana	4	9	0.90	<1	1	0.01
Iowa	635	1400	45.70	270	722	23.55
Kansas	993	2185	59.72	484	1305	35.68
Michigan	121	257	2.70	4	11	0.12
Minnesota	386	845	19.15	509	1366	30.96
Missouri	130	267	4.56	1	1	0.02
Nebraska	992	2182	96.93	228	614	27.26
North Dakota	328	726	24.08	1013	2744	91.02
Ohio	72	148	1.19	1	2	0.02
South Dakota	774	1694	206.31	561	1516	184.64
Wisconsin	137	284	5.82	1	3	0.05
Total for Midwest	4598	10050	13.75	3072	8285	11.34

Source: Brower *et al.* (1993). *Powering the Midwest – Renewable Electricity for the Economy and the Environment.* Cambridge, MA: Union of Concerned Scientists.

generation of machines (Hock *et al.*, 1992) and in future development of Midwest sites. Off-shore locations have already been used along the coasts of Europe (Evans, 1992).

There are other aspects of siting that affect public acceptance of wind farms, which warrant consideration as well. These include land use, visual impacts, noise, communication interference, and bird strikes (Grubb, 1993). Noise and interference can be reduced through technical measures to mostly acceptable levels and are localized in any case to approximately 400 m (1300 ft) from the site (Sesto & Casale, 1994). Such public concerns are mitigated largely by rural siting, where joint use of the land tends to compensate for impacts to those exposed locally to the sites. The proposed use of wind farms in the agricultural lands of the Upper Midwest provides the best example of the expected acceptance there of large wind farms. However, turbine operators must give careful attention to the neatness of operations and local landscapes in order to preserve public support in the surrounding communities (Gipe, 1996).

Bird kills have aroused some controversy (Asmus, 1994), centered on some of the existing California wind farms. The kills, while small in absolute numbers, may be significant for certain species (e.g. eagles). Evidence to date has been fragmentary. Factors specific to the sites and to

Figure 3.6
A small wind turbine from Southwest AIR turbine (300 kW). (Courtesy: American Wind Energy Association.)

the behaviour of the various bird species are poorly understood, leading to calls for more complete studies by the (federal) Fish & Wildlife Service. One such study could compare bird mortality rates (past and projected) linked to wind generation of power with those from other causes, which are known to be in the millions annually. Also, trends toward large turbines, with slower-moving blades, is predicted to reduce the number of strikes (Kerlinger, 1998).

It is well to observe that promising wind sites are not restricted to any one latitude or region, as occurs for solar-energy sites, which are uniformly confined to the sunbelt. It has been estimated in the USA that 90% of the nation's cattle range land, 70% of the agricultural land, and 50% of forest ridges could be available for wind-energy production without hindering their basic uses (Hock, *et al.*, 1992). Greater efforts at data collection for future projects are planned.

Among the wind turbine design developments of the 1980s, smaller machines, with rotor diameters of around 3m (10 ft), also reached the market (see example in Fig. 3.6). They typically had output capacities of around 1 kW and were intended for remote-site, stand-alone use. These machines are designed to be in non-attended, rugged service and to be self protecting in high winds (by tilting to a vertical position) (De Meo & Steitz, 1990). Such units would be appropriate for wilderness sites or remote-farm use, such as in the LDCs. Such applications are comparable with the use of photovoltaics at remote sites (see p. 58).

A final consideration of wind-farm siting is the relationship to the intended load, usually a utility grid. Not only is proximity important (for transmission) but also how time profiles of generation will match the profile of the load demand. For Altamont Pass and the other California wind farms, the diurnal rise of sea-breeze winds, as the sun rises, creates an excellent match with daily electric demand profiles of Pacific Gas & Electric (PG&E) and the other California utilities along the coast (Smith, 1987). A similar coincidence has been forecast for at least one utility in the Upper Midwest, if and when wind generation is installed there (Brower *et al.*, 1993). Whenever such a situation exists for any intermittent generating source, whether wind or solar, a significant fraction of the installed (intermittent) capacity can be given "capacity credit", to be added in with installed conventional capacity toward the utility systems' reliability to deliver power to loads on demand (EPRI, 1990; Kelly & Weinberg, 1993; Cavallo *et al.*, 1993).

For siting prospects where wind patterns might not match load demands, wind offers possibilities in select locations of using pumped-hydro storage of generated wind energy. Such prospects would not be available for solar projects in sunbelt regions but are realistic possibilities in the northwest USA and elsewhere. Pumped-hydro storage has even been proposed to capture some of the wind energy generated during off-peak periods at Altamont Pass, since the PG&E utility already operates such a storage plant (Smith, 1997). Wherever such storage is possible, the intermittency of wind generation becomes less of a disadvantage and the wind-generated power can be considered for a greater firm capacity to the utility system. Until other utility-scale energy storage technologies (see Chapter 5) are available at affordable costs, however, wind generation with storage is limited to a few fortuitous sites.

3.5   Wind technology

Significant technological advances have been made in recent years in electrical generation from wind machines, resulting in improved electrical operation and also gains in aerodynamic energy capture (Dodge & Thresher, 1989; DOE, 1992, 1994; Ramakumar, 1992; NREL, 1992). One of the long-standing problems of wind generation has been the variations in rotational speeds of the rotor, resulting from the ever-present fluctuations in wind speeds. The electric generator of choice, until recently, has been the classic "induction machine" (Fitzgerald, 1983; Heier *et al.*, 1994). Induction machines do not operate exactly synchronously with the utility power system, where the 60 Hz frequency is maintained rather precisely by the system's synchronous generators. An induction machine acting as a motor has a continual slip in its rotational speed below the synchronous speed; however, to act as a generator, it must be driven faster than the synchronous speed.

When driven as a generator, an induction machine delivers power only over a narrow range (10%) of rotational speeds above the 60 Hz synchronous speed, with an even more narrow 2–3% range for maximum efficiency, only rarely maintained by the wind-turbine drive. Aerodynamically speaking, the wind turbines driving the induction generator deliver maximum energy when the blade tip–speed ratio is in its optimum range, as mentioned above. If the wind speed varies, maintaining this range of tip–speed ratios requires that the rotational speed of the rotor must vary along with the wind, subject only to some (not too effective) mechanical or aerodynamic controls. Consequently, the conventional induction machine is non-optimal for wind service on two counts: it does not deliver quality a.c. (synchronous) electricity and it has an inherent design conflict between the a.c. synchronism and maximum wind-energy conversion.

The emergence of a new operational concept – variable speed/constant frequency (VSCF) – has begun to answer some of these difficulties. While initial attempts at VSCF drives mechanical means, such as continuously variable transmissions, the most feasible approach appears to be through "power electronics", working with microelectronics controls (Ramakumar, 1992; EPRI, 1990; Heier *et al.*, 1994). The electronics are fed by a generator with permanent magnet poles and output frequency varying with the rotor speed. The power electronics takes the variable-frequency waveform of the generator and chops it into a 60 Hz (power-line) frequency; control is achieved with a micro processor sensor of the synchronous frequency and the phase of the power system being fed. By these electronic means, a wind rotor with a varying rotational speed can be made to deliver a constant-frequency output synchronized to the power system. Another way to achieve VSCF operation is to rectify the a.c. waveform of a generator which has a variable output frequency and then invert the d.c. voltage back to an a.c. waveform with a controlled constant frequency (Sesto & Casale, 1994). Using such a system, the generator is decoupled from the grid frequency, or excitation fields, and its a.c. output can vary with the wind while synchronism with the grid frequency is maintained by the power electronics.

In addition to the benefits of a synchronized output frequency and better optimized aerodynamic operation, the VSCF system also delivers overall gains in annual outputs of energy and in capacity factors, all of which has a direct impact on the financial returns of wind generation. Furthermore, another problem associated with a.c. power systems – that of the power factor – is also alleviated. The conventionally used induction machines draw "reactive power" (out-of-phase current components), which increases circuit losses and contributes nothing to the useful (watt) output. The reactive power for induction machines can be avoided using capacitors, but these add expense, especially if their use must be controlled. The electronic VSCF systems have no induction machine to draw reactive power and, in addition, the reactive power (and hence the power factor) of the controlled rectifiers are easily and inexpensively controlled.

In summary, it appears that wind-generation technology has a promising future, evidently less problematic in the short to medium term than either form of solar electric generation. Near competitive operation in utility generation has been achieved in several locations in California, Hawaii, and northern Europe, and further penetration into the market continues. Reliable operation has been attained using mid-range capacity machines, operated in large wind farms. All told, these operations total over 16000 individual turbines, with an aggregate capacity more than 1.8 GW in the USA and producing in the 1990s about 3.0 TW-hr annually on a routine basis (Hock *et al.*, 1992). This aggregate wind capacity is about 0.25% of the total installed (700 GW$_e$) conventional capacity in the USA and its energy output is a little over 0.1% of the total US electric output (2700 TW-hr). Meanwhile, even larger wind programs are being pursued in Europe, both in individual countries (e.g. Belgium, Denmark,

Germany, Italy, Spain, and the Netherlands) (Grubb, 1993; Gipe, 1995a; EUREC, 1996) and with the support of the EC (Evans, 1992; Lamarre, 1992). Wind machines built in the USA as well as European machines are being used in these prototype demonstrations (DOE, 1994). The support includes funding for feasibility studies plus production incentives. Over 4 GW_e of wind capacity has already been installed in aggregate within the EC, with a goal of 8 GW wind capacity by the year 2005. The technical potential for wind generation in the EC is many times greater but likely cannot be approached if site restrictions (e.g. urban areas) are applied, as they were for the Upper Midwest in the USA. Significant projects have also been started in Third World countries, such as Bolivia, China, and India (Flavin, 1995; Abramowski & Pasorski, 1998). Worldwide, the aggregate installed wind capacity is approaching 8.0 GW, with growth rates between 1 and 2 GW per year (Gipe, 1995a,b; Brown *et al.*, 1997; Rackshaw, 1998).

Further technology innovations, such as VSCF drives, advanced controls, improved air foils, better wind forecasting, and higher towers promise not only improved technical operation but also economic gains as well (Hock *et al.*, 1992). Further design improvements should expand the range of operable wind speeds, thus increasing annual available hours and capacity factors. Modularity of wind machines operated in wind farms has the same appeal to project developers as solar-trough farms (De meo & Steitz, 1990) Incentives have been necessary for promotion of large-scale projects, but wind farms seem to have survived better after loss of federal tax credits than did the solar-trough projects. The cost goals of the promoting agencies (public and private) is to achieve capacity costs under $1000/kW_e$ and (busbar) cost of energy under $0.05/kW-hr for good wind-resource sites (UWIG, 1992; DOE, 1994; EUREC, 1992). Progress towards this goal as a maturing technology (Weinberg, 1994) is taking place in a steadily evolving manner (Parsons, 1998).

3.6 Market penetration

Even with government support (R&D and financial incentives), the penetration of wind generation into the electric utility market can be expected to be limited. As an intermittent source, there has been a general expectation that supply of electric energy to any utility grid system would be limited in the absence of utility-scale energy storage. It has been estimated by some analysts (Hock *et al.*, 1992; Grubb, 1993) that wind generation will be limited to a maximum of 20% of installed total generating capacity from all sources and its role would be confined to displacing higher-cost fossil fuels for electric generation. There has, however, been somewhat of a debate in recent years on the issue of allowable penetration, as it pertains to both system reliability *and* system economics (Cavallo *et al.*, 1993). Some studies indicate that as much as 45% of the load-energy demand can be supplied in regions of higher wind-class resources, while other analyses on this basis support the less optimistic 10–15% range. Even though wind-energy resources represent over ten times the energy requirements now met by conventional generation in the promising US region of the Upper Midwest, the limitations imposed by intermittency of the wind source could restrict wind generation to approximately 15% of installed capacity in the region. Even this, however, would imply a substantial 6 GW (6000 MW) wind-generation capacity out of a total installed capacity of about 40 GW (40000 MW) for the seven state region (NERC, 1996).

The question of what is the optimum fraction of wind capacity can turn more on the assessment of costs, such as how much wind capacity investments offset the need for investment in conventional generators for "spinning reserves" necessary to preserve the system reliability, rather than on the reliability criteria themselves. However, energy storage, such as batteries, could be considered to improve system reliability by providing "buffering" over short-time (minutes) wind fluctuations (see Chapter 5), thus avoiding reliance on spinning reserves. This, however, would also have added investment costs. Consequently, it is difficult to draw clear conclusions, the situation being dependent on the particular conditions facing a local utility.

A full accounting for the economics of wind installation on utility grids must include the fixed (investment) costs with the variable (operational) costs, subject to constraints of acceptable reliability. The entire mix of generator types, their vintages, and operational records must be considered in analyzing any given power system for capacity additions of any type, conventional or non-conventional (Stoll, 1989). Transmission capabilities and investment costs must also be included. At the present state of the technology, the question to be answered for wind penetration in any given region and power system is: "What beneficial displacement of conventional installed capacity and usage can be made, while preserving the same level of reliability, for least-cost electricity?" Only this can give a credible estimate of "full capacity credit" for intermittent sources such as wind generation.

A feature that will limit the siting of wind generation in particular regions is electric transmission. When the investment costs for transmission lines are considered, the siting of a wind farm any distance more than 100 miles (160 km) from its utility load would be prohibitively expensive, unless possibly the size of the wind project would warrant high-voltage transmission. These are the same kind of cost burden that must be accounted for in siting hydroelectric plants (see Chapter 4). For distances up to 300 miles (500 km), high-voltage a.c. (HVAC) lines would be the most economical, whereas high-voltage d.c. (HVDC) lines are best for more than 300 miles (Wu, 1990). At the present (marginal) level of wind-power capital costs, remote siting is probably excluded. Only if the hoped for reductions in initial costs to well below $1000/kW are achieved would certain remote-site projects become feasible, as they are for some hydro projects. This limitation, therefore, severely restricts the use of the vast wind-resource land areas discussed above until sufficient cost reductions are made for the wind technology itself.

Despite the limitations apparent at this time, the DOE is optimistic about the future market penetration of wind power (Hock *et al.*, 1992; Wald, 1992). Unit capacity costs have dropped to below 50% of those in the 1980s, reaching less than $1000/kW in the 1990s. Over $3B alone has been invested in California wind projects and over $1B is projected for the Upper Midwest. The DOE forecasts a 100-fold increase in wind generation by the year 2030, when they expect wind power to displace 3–4 QUADS (3–4 EJ) primary energy from the national energy budget. At the present time, about 28 QUAD (29 EJ) primary energy is required for conventional electricity generation in the USA, out of a national total around 80 QUAD (84 EJ) for all types of energy. Their forecasts are based on assumptions of the technology gains mentioned above and siting at moderate-to-good wind-resource sites throughout the country, such as the West Coast and the Upper Midwest. The British Wind Energy Association has a target of 6% of Great Britain's electricity from wind, using substantial off-shore siting schemes (Anon., 1998).

3.7   The Wind R&D program

The DOE Wind Energy Program is currently sponsoring R&D for the next-generation turbine technology and promoting new projects with wind machine manufacturers. It has been investing around $50M annually in wind R&D in the mid-1990s; however, this had fallen over 30% by the end of the decade. (DOE, 1995). Wind farm operators and electric utilities relating to this program have set "cost of energy" (COE, see Appendix C) goals for themselves. Earlier goals, using assumptions such as those cited above, were $0.05/kW-hr by the year 2000, all taken at the sites of "moderate to good" wind resources. These goals were upgraded to $0.025–0.035/ kW–hr in 1997 (DOE, 1997) and R&D cost-shared contracts were allowed to $40M to turbine manufacturers to help to achieve them. Currently, wind generation has established the lowest cost of energy of the newly developed renewables, such as solar–thermal and solar–PV, thus giving more cause for such optimistic projections than for other technologies. Experience in Northern Europe (Gipe, 1995b) suggests that market pull can stimulate further investment in wind generation (see Chapter 11 for further discussion).

Wind energy is a prime example of a *sustainable* resource: it is non-polluting, inexhaustible, and has no emissions or residues to burden society. While limited by site and, for the time being, by its intermittency, it stands out as a competitor to fossil fuels, with early successes. It also stands out as an example of the value of public incentives in aiding early penetration of a new technology into the market. Wind farm operators were able to achieve viability, using tax incentives before the incentive programs were cut back and now can produce energy at a competitive cost in several regions throughout the world. Attempts to demonstrate and commercialize other renewables, such as the solar-thermal (trough collector) program, were not completed before cutbacks occurred in the incentive programs, even in ideal (desert) locations.

References

Abramowski, J. & Posorski, R. (1998). Wind energy – a true option or developing countries. *Renewable Energy World*, 1, 62–64.

Anon. (1997). Enron wins pact to supply power from wind turbines (Business Section). *New York Times,* 20 March, D2.

Anon. (1998). UK forsees boom in wind power schemes. *Oil & Gas Journal*, 16 March, 49.

Asmus, P. (1994). Hot air, hot tempers and cold cash – clashes of ethics and clashes of interests. *Amicus Journal*, Fall, 30–35.

Beyer, H.G. Luther, J. & Steinberger-Willins, R. (1993). Power fluctuations in spatially dispersed wind turbine systems. *Solar Energy*, 5, 297–305.

Boer, K.W. (ed.) (1994). *Advances in Solar Energy,* Vol. 9, Boulder, CO: American Solar Energy Society. (a) Sesto, E. & Casale, C. Wind power systems for power utility grid connection, Ch. 2, 71–159; (b) Heier, S. Kleinkauf, W. & Sachau, J. Power conditioning, Ch. 3, 161–243.

Brower, M.C., Tennis, M.W., Denzler, E.W. & Kaplan, M.M. (eds.) (1993). *Powering the Midwest – Renewable Electricity for the Economy and the Environment.* Cambridge, MA: Union of Concerned. Scientists.

Brown, L.R., Renner, M. & Flavin, C. (1997). *Vital signs – 1997. The environmental trends that are shaping our future.* New York: Norton for World Watch Institute.

Cavallo, A.J. Hock, S.M. & Smith, D.R. (1993). Wind energy: technology and economics. In: *Renewable Energy – Sources for Fuels and Electricity*, Ch. 3, eds. Johansson, T.B., Kelly, H., Reddy, A.K.N. & Williams, R.H., pp. 121–156. Washington, DC: Island Press.

De Meo, E.A. & Steitz, P. (1990). Wind power. In: *Advances in Solar Energy*, Vol. 6, ed. Boer, K.W. Boulder, CO: American Solar Energy Society.

Dodge, D.W. & Thresher, R.W. (1989). Wind technology today. In: *Advances in Solar Energy,* Vol. 5, Ch. 4, ed. Boer, K.W. Boulder, CO: American Solar Energy Society.

DOE (US Department of Energy) (1992). *Wind Energy Program Overview, Programs in Utility Technologies, Fiscal Years* 1990–91. DOE/CH 10093–101. Golden, CO: National Renewable Energy Laboratory.

DOE (1994). *Wind Energy Program Overview – Fiscal Year 1993.* DOE/CH10093–279. Golden, CO: National Renewable Energy Laboratory.

DOE (1995). *Task Force on Strategic Energy Research and Development, Annex 1: Technology Profiles*, Washington, DC: Secretary of Energy Advisory Board

DOE (1997). *Next Generation Turbine Development.* Golden, CO: National Renewable Energy Laboratory. (As reported in *Solar Today*, Sept./Oct. 41–42, (1997).)

Dunn, P.D. (1986). *Renewable Energies: Sources, Conversion and Application.* London: Peregrinus for IEE.

Eldridge, F.R. (1980). *Wind Machines.* New York: Van Nostrand.

Elliott, D.L. & Schwartz, M.N. (1993). *Wind Energy Potential in the United States.* National Wind Technology Center – NREL/USDOE < http://nwtc.nrel.gov/publish_papers/potential.html >

EPRI (Electric Power Research Institute) (1990). Wind energy coming of age. *EPRI Journal*, June, 15–25.

EUREC (EUREC Agency) (1996). *The Future for Renewable Energy – Prospects and Directions*. London: James & James.

Evans, L.C. (1992). Wind energy in Europe, *Solar Today*, May/June, 32–34.

Fitzgerald, A.E. *et al.*, (1983). *Electric Machinery*, 4th edn, Ch. 9. New York: McGraw Hill.

Flavin, C. (1995). Harnessing the sun and wind. In: *State of the World (1995)*, Ch. 4. New York: Norton for World Watch Institute.

Frankel, E. (1986). Technology, politics and ideology: the vicissitudes of federal solar energy policy, 1974–1983. In: *Energy Policy Studies*, Vol. 3, eds. Byrne, J. & Rich, D. New Brunswick, NJ: Transaction Books for the University of Delaware.

Gipe, P. (1995a). Wind energy's declining costs. *Solar Today*, Nov./Dec., 22–25.

Gipe, P. (1995b). Wind comes of age. New York: Wiley.

Gipe, P. (1996). Cleanup time (Letter). *Solar Today*, Nov./Dec., 6–7.

Grubb, M.J. (1993). Wind energy, In: *Emerging Energy Technologies: Impacts and Policy Implications*, eds. Grubb, J., Walker, J., Ruxton, R. *et al.*, Ch. 10. Dartmouth, UK: Dartmouth Publishing for Royal Institute of International Affairs.

Grubb, M.J. & Meyer, N.E. (1993). Wind energy: resources, systems and regional strategies. In: *Renewable Energy – Sources for Fuels and Electricity*, Ch. 4, eds. Johansson, T.B., Kelly, H., Reddy, A.K.N. & Williams, R.H., pp. 157–212. Washington, DC: Island Press.

Heier, S. Kleinkauf, W. & Sachau, J. (1994). Power conditioning. In: *Advances in Solar Energy*, Vol. 9, Ch. 3, pp. 161–243. ed. Boer, K.W. Boulder, CO: American Solar Energy Society.

Hickok, F. (1975). *Handbook of Solar and Wind Energy*. Boston, MA: Cahners Books International.

Hills, R.L. (1994). *Power from Wind: A History of Windmill Technology*. Cambridge: Cambridge University Press.

Hock. S., Thresher, R. & Williams, T. (1992). The future of utility-scale wind power. In: *Advances in Solar Energy*, Vol. 7, Ch. 9, ed. Boer, K.W., pp. 309–371. Boulder, CO: American Solar Energy Society.

Kealy, E.J. (1987). *Harvesting the Air: Windmill Pioneers in 12th Century England.* Berkley, CA; University of California Press.

Kelly, H. & Weinberg, C.J. (1993). Utility strategies for using renewables. In: *Renewable Energy – Sources for Fuels and Electricity*, Ch. 23, eds. Johansson, T.B., Kelly, H., Reddy, A.K.N. & Williams, R.H., pp. 1011–1069. Washington, DC: Island Press.

Kerlinger, P. (1998). Wind turbines and bird safety. *Renewable Energy World*, Nov., 7.

Lamarre, L. (1992). A growth market in wind power. *EPRI Journal*, Dec., 14–15.

NERC (National Electric Reliability Council) (1996). *Regional Assessments – 1996*. Princeton, N J: National Electric Reliability Council.

NERL (1992). *Cooperative Field Test Program, NREL/TP 253-4252*. Golden, CO: National Renewable Energy Laboratory.

Parsons, B. (1998). Grid-connected wind energy technology: progress and prospects. *Newsletters, 4th Quarter.* Cleveland, OH: International Association of Energy Economists.

Rackshaw, K. (1998). Wind around the world. *Solar Today*, March/April, 22–25.

Ramakumar, R. (1992). Wind energy systems. Report on 1992 Summer Meeting Panel Session on developments in wind–electric conversion system technology. *IEEE Power Engineering Review*, Dec., 6–9.

Sesto, E. & Casale, C. (1994). Wind power systems for power utility grid connection. In: *Advances in Solar Energy,* Vol. 9, Ch. 2, ed. Boer, K.W., pp. 71–159. Boulder, CO: American Solar Energy Society.

Smith, D.R. (1987). The wind farms of the Altamont Pass. *Annual Review of Energy*, 12, 145–183.

Stoll, (1989). *Least-Cost Electric Utility Planning.* New York: Wiley.

Swisher, R. & Gipe, P. (1992). US windfarms: an expanding market. *Solar Today*, Nov./Dec., 17–19.

UWIG (Utility Wind Interest Group) (1992). *America Takes Stock of a Vast Energy Resource* (Brochure, Feb). Golden, CO: National Renewable Energy Laboratory and Palo Alto, CA: Electric Power Research Institute.

Wager, J.S. (1994). Renewables in the MidWest. *Solar Today*, March/April, 16–18.

Wald, M.L. (1992). A new era for windmill power, *New York Times,* Business Day, 8 Sept.

Weinberg, C.J. (1994). The electric utility; restructuring and solar technologies, Chapter 5, In: *Advances in Solar Energy,* Vol. 9, Ch. 1, ed. Boer, K.W. Boulder, CO: American Solar Energy Society.

Williams, S. & Bateman B.G. (1995). *Power Plays – Profiles of America's Independent Renewable Electricity Developers.* Washington, DC: Investor Responsibility Research Center.

Wu, C.T. (ed.) (1990). AC–DC Economics and Alternatives – 1987. *IEEE Transactions on Power Delivery, Panel Session Report*, Vol. 5, 1956–1976.

Further reading

Grubb, M.J. (1991). Value of variable sources on power systems. *IEE Proceedings*, 138, 149–165.

Jayadev, J. (1995). Harnessing the wind. *IEEE Spectrum*, Nov., 78–83.

Johansson, T.B., Kelly, H., Reddy, A.K.N. & Williams, R.H. (eds.)(1993). *Renewable Energy – Sources for Fuels and Electricity.* Washington, DC: Island Press. (a) Cavallo, A.J., Hock, S.M. & Smith, D.R. Wind energy: technology and economics, Ch. 3, pp. 121–156; (b) Grubb, M.J. & Meyer, N.E. Wind energy: resources, systems and regional strategies, Ch. 4, pp. 157–212.

Johnson, G.L. (1985). *Wind Energy Systems.* Englewood Cliffs, NJ: PrenticeHall.

Twidell, R. (1987). *A Guide to Small Wind Machine Energy Conversion.* Cambridge: Cambridge University Press.

USGAO (US General Accounting Office) (1993). Electricity supply – effects under way to develop solar and wind energy. USGAO Report to the Chairman, Subcommittee on Investigations & Oversight, Committee on Science, Space & Technology, *US House of Representatives, Report No. GAO/RCED-93-118*, April. Washington, DC: US General Accounting Office.

4 Hydroelectric power

4.1 Introduction

Hydro power is the premier renewable energy resource, its very nature exemplifying the definition of renewable. In addition, it is non-polluting and emits no carbon dioxide in operation. These attributes long predated the concept of sustainability but obviously fit it in an essential way. However, hydroelectric generation is not a new technology in need of R&D to prove its technical viability. We have no need here for such assessments, as we have done for the new prospective technologies elsewhere in this book. Other aspects of hydropower do call for assessment, including its environmental and social impacts.

The use of water power goes back to antiquity, with the use of water wheels turned either by the flow of a river (or stream) or by the weight of water caught falling at a dam or waterfall. The sites where this source of mechanical power was available determined the locations of mills and, later, factories in the old world and the new (Hunter, 1979; Golof & Brus, 1993). The invention of the electric generator and the hydraulic turbine in the late 19th century provided the means to exploit hydro resources through electricity generation (Dowling, 1991; Moreira & Poole, 1993). One of the original turbines, the Francis type, was designed and used in the mid-19th century and some of its descendants are still in use today.

Our assessments of hydro will concentrate mostly on its environmental and social impacts followed by an assessment of what contribution it can make to the world's sustainable energy needs in view of those impacts. Most concerns over adverse impacts have been centered on large (multi-gigawatt electric power) projects, whereas small projects (less than 30 MW$_e$) have been viewed generally in a positive way (Deudney, 1981; Golof & Brus, 1993). Nonetheless, there are important variations within each category, depending on the particular project, its site, and planning (Goodman, 1984; Moreira & Poole, 1993).

The present use of hydroelectricity is already significant, with about 20% of the world's total annual electricity production (12 000 TW-hr) being water driven. For a number of countries, hydro is the major electricity source, including Canada and Austria with over two thirds of capacity being hydro, Brazil with close to 90%, and Norway and Zambia both with close to 100% (Deudney, 1981). More hydro projects are under construction or in planning currently, which will add measurably to the global generating capacity.

The potential for increased hydro generation is dependent on the world's water balance and the net annual runoff of rain from land areas to

Table 4.1. *Hydroelectric generating capacities*

Region	Technically exploitable potential (GW$_e$)	Already exploited generating capacity (GW$_e$)	Exploited (%)
Asia	610	98	16
South Asia		45	
China		33	
Japan		20	
Latin America	432	96	22
South America		85	
Central America		11	
Africa	358	17	5
North America	356	148	42
Canada		58	
United States		90	
Former USSR	250	62	25
Europe	163	145	89
Eastern Europe		17	
Western Europe		128	
Oceania	45	12	27
World total	2214	577	25

Adapted from Deudney, (1981), *IWPD* (1989), Brown *et al.* (1993), Moreira & Poole (1993).

the oceans. Were we to estimate the hydro potential simply from runoff, Asia has the largest of any land mass, with 30% of the world total. South America has nearly as much (26%), with less than half the land area, and North America has only about 17% of the world's runoff. The estimated potential generating capability for hydro in a given region depends not just on precipitation runoff but also on the land topology, civil construction requirements, and environmental constraints. The "technically exploitable potential" is defined in a manner that includes hydrologic and engineering factors at potential hydro sites but does not exclude sites on the basis of complete environmental and social impact analyses. These potentials are generally given as the aggregate generating capability of all feasible hydro sites for a given region, in terms either of rated electric output power capacity (GW$_e$) or annual electric energy output capability (TW-hr).

The technically exploitable potentials for the various regions of the world are shown in Table 4.1 in terms of potential generating capacities (in GW$_e$ of electric power). Also shown are the amounts of these potentials actually exploited (installed and operating capacities) as of 1992. It is to be noted that the developed countries have utilized the higher fractions of exploited hydro potentials, whereas the underdeveloped countries and the former USSR have much lower proportions utilized. In the USA, for example, there is about 90 GW$_e$ hydro capacity, which represents around 40% of the potential. In Europe, almost 90% is utilized, whereas in Asia the figure is only 16% and it is only 5% on the African continent.

If more detailed analyses are made of the environmental impacts and the costs of construction evaluated for potential sites, the so-called "economically exploitable potentials" can be estimated for the various regions. The world totals on this basis have been judged likely to be 40–60% of the technically exploitable aggregates (Moreira & Poole, 1993), but this would still allow for an expansion of world hydro capacity to over three times its present level. Further consideration of social impacts would no doubt reduce these potentials even more through either accounting for the costs of social externalities or out-and-out exclusion of certain projects because of unacceptable social impacts. Nonetheless, significant expansion of hydroelectric generation appears feasible on an environmentally and socially sustainable basis, given adequate care in planning.

4.2   Impacts of hydro projects

It is the large hydro projects that have provoked the most criticism on environmental and social grounds (Deudney, 1981; Moreira & Poole, 1993). Large hydro projects inevitably require large dams and reservoirs, which inundate large land areas. An example is the huge Itaipú project in South America, shown in Figure 4.1a. This dam, located near the border between Brazil and Paraquay, is 150 m (490 ft) high and the reservoir is 8 km (5 miles) long. Such reservoirs typically have capacities of approximately 25 km^3 (20 million acre-ft). Large hydro generation capacities are in the multigigawatt electric power category, requiring large high-voltage transmission lines to send the power to remote load centers. The construction projects themselves are disruptive to the surrounding area because of the shear size of the multibillion dollar activities employing tens of thousands of workers. Public-health impacts can occur through boom-town conditions or from water-borne diseases such as malaria.

The land inundations are large, some running to hundreds of thousands of hectares (approaching a million acres), often displacing large populations of people, including indigenous peoples (Park, 1992). Invariably, wildlife habitats are disrupted to some degree (Raphals, 1992). The shear masses of water in larger reservoirs have even induced seismic activity in some areas. Aquatic ecosystems can be disrupted by eutrophication in reservoirs of former vegetated land and the oxygen content of discharged water downstream from large dams has often been found to be too low for fish to survive. In addition, changed sedimentation patterns along river paths can be detrimental, sometimes to the dam projects themselves. Figure 4.1b gives a summary of hydropower impacts.

Most of these hydro impact problems have been addressed and many mitigated with measurable success. The key to effectiveness comes with prior planning for the social–environmental dimensions of the project, both in construction and in subsequent operation (Moreira & Poole,

1993). The principles and procedures for such planning are well developed (Goodman, 1984; Shaw, 1993). All aspects, including wildlife, fisheries, watershed, population displacements, and associated irrigation projects, have to be evaluated in preplanning stages before decisions to proceed with the projects are made. Scenic, aesthetic, and recreational valuations can be carried out as part of the integrated planning, and indeed have been statute requirements in some countries (e.g. exclusions of hydro in the US *Scenic Rivers Act*). In some cases, as a result of preplanning screening, it will turn out that the project is not feasible, either because of unacceptable environmental or social impacts or because the mitigation of the externalities makes the project economically unfeasible. When any such project is excluded, its potential power-producing capacity is taken out of the aggregate, economically exploitable potential for the region.

Public involvement in the decision process for hydro projects is essential, not only out of democratic principles but also because it benefits the planning process itself (Goodman, 1984). The determination of what are acceptable impacts, or what is sufficient mitigation to allow a large project to go forward, are matters for public valuation going beyond the cost–benefit studies traditionally used for dams and irrigation projects (Thompson, 1980; Fischhoff & Cox, 1986; Fischer & Forester, 1987). Expectations have grown in democratic countries since the 1960s for opening the decision processes for such large projects to public comment and intervention. Environmentalists and advocates for indigenous peoples have vigorously opposed projects such as the James Bay complex in Quebec (Bourassa, 1985; *Globe & Mail,* 1991–92; Verhovek, 1992) and the large hydro projects in Brazil (Moreira & Poole, 1993). These actions have no doubt heightened awareness and sensitivities to these issues; as a result further limitations on hydro exploitability are likely as time progresses.

4.3 Small-scale hydro projects

Small-scale hydro projects do not have the impacts of the large plants, either environmental or social. Generating capacities of small hydro projects are in the megawatt range (1–30 MW$_e$), including subcategories of "mini-hydro" and "micro-hydro", which are less than 1 MW$_e$ and 100 kW$_e$, respectively (Fritz, 1984). Worldwide, an estimated 23.5 GW$_e$ aggregate installed small-hydro capacity existed in 1989 (Moreira & Poole, 1993), which is less than 5% of the total world hydro capacity.

Such plants typically are "run-of-river" plants, working on the flow of water and have little or no dam storage capacity. The impacts of reservoir inundations, river flow, and alike are, therefore, lowered, even considering the reduced scale of these projects. Land area requirements per unit of energy production capacity are proportionally less for small hydro than for large projects, sometimes less than one tenth per unit compared with large-scale projects (Kozloff & Dower, 1993). However, because of the lack of reservoir storage capacities, these plants are not able to produce electricity during dry seasons or unusually dry years, often resulting in low

(a)

Figure 4.1
Hydroelectric projects.
(a) A large project,
Itaipú, Brazil (12.6 GW).
(Courtesy of Morrison
Knudsen International.)
(b) Environmental
impacts of
hydroelectric projects.
(Reprinted with
permission from Lagler,
(1969), *Man-made
Lakes*, Food &
Agricultural
Organization of the
United Nations, Rome.)

CFs and consequent poor use of capital investments. For example, CFs averaging about 33% for some of China's many small hydro units (Deudney, 1981) would result in unacceptably high fixed (investment) costs in a market society.

The small-scale hydro generation plants of China are, nonetheless, impressive examples of the resource possibilities of the technology. Over 100 000 units have been built since 1950, with a total over 9 GW_e aggregate capacity now in operation, generating an aggregate annual energy over 36 TW-hrs (Fritz, 1984; IWPD, 1990). Since the CFs of some of the small plants run are less than 10%, they must be used in grid systems with other types of plant, including large hydro and thermal plants, to maintain the availability of supply, in much the same way that wind and solar generation must be operated. The Chinese have not, therefore, relied solely on small hydro, which represents only about 37% of their total hydro capacity. Construction has commenced (Anon., 1993) on what will be the world's largest hydro plant (17.7 GW_e), at the Three Gorges on the Yangtze River.

In the Western developed countries, small-scale hydro is currently about the only form of further exploitation of hydro potential taking place. In the USA, for example, there were 59 new small hydro plants under construction in 1991, which will add 71 MW_e capacity to the existing 3.42 GW_e

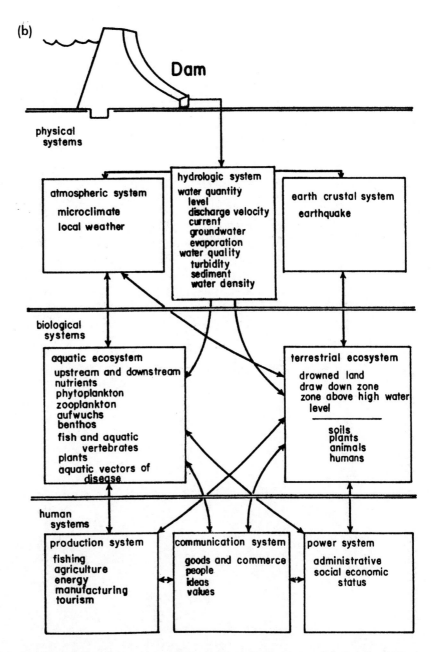

(b)

Dam

physical
systems

hydrologic system
water quantity
level
discharge velocity
current
groundwater
evaporation
water quality
turbidity
sediment
water density

atmospheric system
microclimate
local weather

earth crustal system
earthquake

biological
systems

aquatic ecosystem
upstream and downstream
nutrients
phytoplankton
zooplankton
aufwuchs
benthos
fish and aquatic
vertebrates
plants
aquatic vectors of
disease

terrestrial ecosystem
drowned land
draw down zone
zone above high water
level

soils
plants
animals
humans

human
systems

production system
fishing
agriculture
energy
manufacturing
tourism

communication system
goods and commerce
people
ideas
values

power system
administrative
social economic
status

(3420 MW$_e$) small hydro capacity (IWPD, 1991; Moreira & Poole, 1993). Over 3000 new plants are also in the planning stage. The potential (technically) exploitable small hydro in the USA has been estimated to be at least 2 GW$_e$. Such plants can be expected to have production costs in the range \$0.07–0.10/kW-hr (Williams & Bateman, 1995). Long-standing plants, having covered their investment costs, have costs as low as \$0.03/kW-hr.

In addition, in the USA, older small hydro plants – long out of service – are being rehabilitated and brought on line at a fraction of the investment cost of new plants (Casey, 1993). There are nearly 50 000 candidate small dam sites in the USA (Fritz, 1984), that had been discarded in the era of cheap fuels. The U.S. Department of Energy instituted the National Small Hydropower Program in 1977, later to be coordinated with the *Public Utilities Regulatory Act* of 1978, to promote rehabilitation of old dams and construction of new ones for small hydro (DOE, 1985). An example of one of these projects is shown in Figure 4.2, a run-of-river plant in Maine. In 1988, however, all tax credits for hydro were ended and hydro producers were excluded from the 1.5¢/kW-hr production credit allowed other renewable energy producers (Williams & Bateman, 1995).

All in all, new and rehabilitated small hydro plants can add, in the near term, over 5 GW$_e$ to the present national hydro capacity of about 90 GW$_e$ in the USA(IWPD, 1989). (The total installed capacity of all types in the USA is close to 800GW$_e$.) At the same time, however, re-licensing of existing hydro plants is subject to strict rules for wildlife and environmental protection, which will require additional investments by some utilities to keep the plants going (IEEE, 1990). Nonetheless, it should be recognized that licensing and financing are the only causes of uncertainty here, unlike all of those that accompany the developing technologies.

4.4 Hydroelectric technology

The technology of generating electricity by hydraulic means is, as mentioned above, a mature one. Hydraulic turbines were developed in the 19th century and are of two main types: reaction turbines and impulse turbines (Dowling, 1991). Reaction turbines operate from the flow of water on the

Figure 4.2
A small hydropower project: Upper Barker hydro project in Maine. (Courtesy of CHI Energy Inc., Stamford, CN.)

turbine blades, with relatively low pressures, and are typically used with the lower dams, run-of-river plants. The original Francis-type reaction turbine has been supplemented, starting earlier in the 20th century, by the Kaplan and bulb-type turbines. The Kaplan type is used for the higher head dams (up to 200 m or 656 ft) and the bulb type for the lower heads (less than 50 m). Figure 4.3 shows the blade assembly of a Kaplan type turbine, the impeller blades at the bottom drive the (vertical) shaft to turn the generator rotor. In Figure 4.4, a schematic view of reaction-turbine operation on the flow of water can be seen with a bulb-type turbine.

Impulse turbines operate at the high pressures resulting from high-head (greater than 300 m) dams. The high hydraulic pressure creates a high-velocity jet of water that transfers its impulse momentum to the cup-shaped blades of the turbine (Fig. 4.5). The Pelton impulse-type turbine came into use during the 1930s with the use of high dams such as the

Figure 4.3
A Kaplan reaction-type hydraulic turbine. (Courtesy of Voith Hydro.)

Hoover Dam. Major innovations to either type of turbine seem unlikely and are not essential for further development of the potential of hydro generation.

Advances have been made in recent years in civil construction methods, resulting in considerable savings in the investment costs of hydro projects (Moreira & Poole, 1993). Roller compacted concrete (RCC) uses special cement mixtures that can be moved into place and compacted to shape with ordinary earth-moving equipment, rather than pouring in liquid form to specially constructed molds. The result is a much speedier construction, with less cement required in the mixture; this brings costs for concrete construction down to one third that of the conventional concrete methods. Savings such as these contribute to making prospective projects economically feasible at more sites and, therefore, to making more of the potential resources available.

4.5   Remote-site transmission

The electric generators used for hydro power are also a mature technology, having been developed earlier in the 20th century. Whereas innovations in the generators are also not vital to further exploitation of hydro, the means of electric transmission is. Hydro sites are typically located in regions remote from their urban or industrial loads and must be connected by long-distance, high-voltage transmission lines (IEEE, 1992). Prime examples are the interconnections from the James Bay complex to southern Quebec and the New England urban centers (a 600 mile (960 km) distance) and the interconnection of the hydro resources in the Amazon River basin of Brazil to cities to the east and south (over 1300 miles or 2100 km).

Transmission line investment costs are a major determinant in the economic viability of most hydro sites. The high-voltage, transmission-line technology is key to lower costs and lower operating losses (EPRI, 1982). It is also central to enhancing the availability of power to electric-grid networks and allowing possible "complementarity" of hydro sources. Hydro plants can have diverse patterns of hydrological runoffs, because they are located in regions remote from one another. Complementarity means that the plants can fill in for one another, since the dry season for one will be the wet season for the other. An extensive proposal for using hydrologic complementary diversity has been made for South America by Moreira and Poole (1993). This would interconnect hydro power north of the equator with that to the south, calling it an "inter-hemispheric interconnection". The highest river flows at the northern tip of the continent (Venezuela, Columbia, etc.) occur in May to August, whereas the highest flows occur in the south (Brazil, Peru, etc.) in January to March. It is speculated that interconnection across the equator may be feasible on the African continent also.

A massive hydro project, with interconnections that would be interhemispheric, has been proposed by the United Nations for the Zaire River

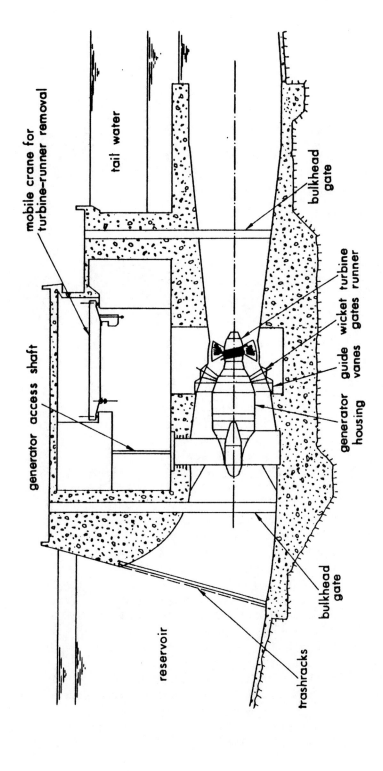

Figure 4.4

A bulb-type hydraulic turbine. (Reprinted with the permission of the publisher and the author of *Energy: Non-nuclearTechnologies*, Vol. II, S.S. Penner, 1975, Addison-Wesley Publishing Co., Reading, MA.)

Figure 4.5
A Pelton impulse-type
turbine.(Courtesy of
Voith Hydro.)

in Central Africa (IEEE, 1992). The generating complex would be called
the Grand Inga Project and has a potential of 20 GW$_e$ capacity, which is
claimed would be continuously available. The proposal calls for transmission to North Africa over a distance of about 4000 miles (6500 km)
with interconnections at one or more points to the electric grid of the EC.
The project would deliver about two-thirds of its power to Europe through
interconnections in Turkey, Tunis, or Morocco, and the remainder would
be distributed elsewhere on the African continent. This grandiose project,
as proposed by the UN/DESD Energy Branch, would be completed in
three phases by the year 2012. It would require a total investment of about
$60 B and an annual revenue requirement of about $5.6 B. Financing
would presumably be through international sponsorship involving the
World Bank and the planners claim to be able to deliver electric energy to
the European interconnects at $0.059/kW-hr. If achieved, this would be
nearly 30% less than the cost of nuclear power generated in Europe. The
proposal is of interest here not so much for assessing its feasibility, or even
its desirability, but because it gives such a clear picture, in its extremes, of
the role of transmission with large hydro generation.

Over half of the investment costs for the Grand Inga Project would be
for transmission equipment, such as the (overhead) transmission lines,
(underwater) cable, and terminal equipment. Long-distance electric transmission requires high voltages, obtained by electrical transformers, in
order to limit resistance losses from current flow. As longer distances have
been attempted, higher voltages have been utilized, with voltages over 1000
kilovolts (kV) achieved in the latest UHV (ultra high voltage) technology.

In the Russian power grid, extending thousands of miles across the continent, voltages up to 1150 kV have been used on HVAC transmission lines, which are up to 3000 miles in length. HVDC transmission lines have become the favored technology for very long distances (Nozari & Patel, 1988). For most terrains, HVDC transmission lines have lower costs per mile of length. (One easily recognizable factor in this is that HVDC lines require only two conductors compared with the three required for three-phase HVAC lines.) However, the HVDC terminal equipment (e.g. a.c./d.c. converters and controllers) is more expensive than that for the HVAC lines. The usual outcome of these competing factors is that HVDC is more economic beyond some threshold distances, often taken to be around 300 miles (500 km). Obviously, for transmission distances of thousands of miles, HVDC lines would be expected to be the least expensive. Also long HVAC lines, with other a.c. generators on the interconnect, have undesirable (potentially unstable) dynamic characteristics.

The evolution of (solid-state) silicon-controlled rectifiers (SCRs), operating at high voltages and handling large currents, has made the HVDC technology operable (Nozari & Patel, 1988). Higher voltages and longer transmission distances have been achieved as a result of developments in SCRs over recent years. Some of the most significant technology advances were made in the former Soviet Union, with 1500 kV HVDC lines extending up to about 4000 miles in length. HVDC lines at 1600 kV ($\pm$ 800 kV on each of the two conductors) have been proposed for the Grand Inga lines, each line capable of transmitting 3.5 GW_e (IEEE, 1992). The total transmission losses for the Grand Inga to Europe interconnects, as planned, should be around 22% of the power transmitted from Zaire. This is, however, a higher fraction loss than usual for conventional power grids.

The Grand Inga Project is a clear example of the economics of hydro power, if for no reason other than its scale. The unit capital cost, for the generation excluding transmission, is optimistically estimated at $780/$kW_e$. If true, this is below the customary range of costs $1000–2000/$kW_e$ experienced for hydro projects in the USA and elsewhere. Such a capital cost, as represented, would be 50% or less of those of fossil or nuclear plants and, therefore, overwhelmingly competitive with them, since hydro has no fuel costs. It should be noted, however, that fossil and nuclear plants capital costs reflect the additions to control environmental factors but *no* account has been given by the promoters for Grand Inga's impacts and consequent externalities (IEEE, 1992).

In comparing hydro and thermal plant costs, the assumption is that the hydro plant is sited remotely while the thermal plant is located in the general locality of the urban or industrial load. For Grand Inga, albeit an extreme example, the transmission costs have more than doubled the investment costs of the generation plants. The cost of hydro power delivered to load centers in Europe, via the long transmission lines, would have to compete against power generated by thermal plants in Europe. On a somewhat less grandiose but still impressive scale, the estimated transmission costs for projects planned in Brazil are about one third of the total costs of $0.033–0.039/kW-hr of electricity *delivered* (IEEE, 1992). The distances in these cases (from the Amazon River sites to industrial centers to the east) are in the range 1300–1700 miles (2100–2800 km), which are less than half those of the Grand Inga project.

A key path to enhanced economic competitiveness and widened number of economically feasible sites for hydro, regardless of the scale of any one project, is the further development of the HVDC technology (Hingorani, 1988; Nozari & Patel, 1988). The heart of the a.c. to d.c. conversion process at high voltages is the SCR or thyristor. The SCR-thyristor is a solid-state (silicon semiconductor) controllable rectifier capable of handling large currents and withstanding very high voltages. Since its initial development in the 1950s, great strides have been made in its performance characteristics. Current carrying capacities of individual thyristor packages have been increased from 50 A to 3000 A and the voltages the package will withstand have gone up from 200 V to 6 kV. The SCR packages are currently used in series and parallel combinations for power-electronic circuits that operate at voltages of hundreds of kilovolts and handle hundreds of megawatts of d.c. power. The trend toward higher ratings is expected to continue, which will reduce unit costs and enhance power-handling capacities. In addition, alternative operational techniques in the receiving-end controls and reduction of harmonic frequencies offer possibilities for cost reductions in HVDC systems (Wu, 1990).

These issues of long-distance transmission, which have been important for many years for hydro power, can also arise for the other renewable technologies such as solar and wind, where feasible operation is site specific and remote from load centers. For example, it has been suggested (Kern, 1993) that solar–PV generation in the sunbelt of the USA could conceivably deliver electricity to the Northeast or the Pacific Northwest at net savings over local PV generation in those less sunny regions. Such schemes, however, would require significant reductions in PV costs (see p. 58) to be competitive with conventional sources in those northern sections.

4.6   The future of hydro power

The major emphasis of this book is on the prospects for *new* energy sources that meet the criteria of sustainability. We have reviewed in this chapter the prospects for hydroelectricity and found limitations on the further development of its renewable resources not as a consequence of failings in technological innovation but rather because of the limits of availability of hydrological resources and the constraints on developing them. In the developed countries, these limitations have already resulted in a leveling off of further hydro exploitation. In the USA, for example, hydro capacity will probably not increase over 10% in the first few decades of the 21st century. It is, therefore, useful to our perspective overall to review what the potential contributions of the newly developing renewable resources, such as wind and solar generation, can be relative to hydro.

Hydroelectricity presently contributes the equivalent of about 3 QUAD of primary energy to US energy consumption (Dowling, 1991). This is about 4% of the total primary-energy consumption, most of which is derived from fossil fuels. In terms of electric energy production, hydro is currently producing close to 10% of the total US electricity (2700 TW-hr total in 1990) Similarly, hydro is contributing about 19% of electricity in OECD Europe (out of nearly 2500 TW-hr), 14% in Eastern Europe (out of nearly 2000 TW-hr), and around 15% in the rest of the world (out of about 4000 TW-hr).

In 1992, wind generation in the USA totaled about 2.7 TW-hr electric energy (Brown, *et al.*, 1993), which is about 1% of hydro's electric output for that year, or 0.1% of the total generation. In terms of aggregate electric installed generating capacities in the USA, wind generation totaled about 1.7 GW_e in 1992, compared with hydro's more than 75 GW_e national total in that year. It has been projected, however, that wind generation could grow 100-fold by the year 2030 (Hock *et al.*, 1992), which would make its contribution comparable to hydro. A similar projection has been made for PV generation in the USA for the year 2010 (DOE, 1991). On these projections the prospective energy contributions of these three renewable sources to US energy demands in the first two to three decades of the 21st century are comparable and could conceivably result in close to one quarter of the annual demand for electric energy during that time span. In making such a comparison, however, we should recognize that the projections for hydro production are firm, whereas those for wind and solar carry the uncertainties of new technology discussed elsewhere in this volume.

The technical potential for hydro development is much greater for the developing world and the promised impacts are of much greater importance to the evolving economies of those countries. This is evident from Tables 4.1 and 4.2, which show, respectively, the percent-exploited generating (output power) capacity and the annual electric energy generation for the various regions of the world. We see that Asia, Latin America, and Africa are presently producing 10% or less of the exploitable hydro energy (Table 4.2) resources of their regions, while Europe and North America have exploited over 45% of theirs. Especially striking in this regard is the former USSR, which currently exploits only 6% of its potential hydro annual energy.

Some of the greatest strides to hydro exploitation in the developing world are taking place in South America. Brazil, in particular (IEEE, 1992; Moreira, Poole, 1993), has about a third of South America's potential and has developed close to one-fifth of it. The Itaipú project (Figure 4.1), with a record installed generating capacity of 12.6 GW_e, is a massive example of earlier construction and impacts. Overall, hydro generation in Brazil is currently over 200 TW-hr annually, which approaches the hydro output of the USA (with hydro programs started over a half century ago). Ambitious plans call for expanding Brazil's current hydro capacity by 50 GW_e to about 140 GW_e by the year 2010, attempting to mitigate the impacts by imposing stricter guidelines in the process.

It should be recognized that detrimental environmental and social impacts, of the types mentioned above, have occurred with hydro projects elsewhere. A continuation of these missteps in the Third World, in its

Table 4.2. *Hydroelectric technically exploitable annual energy (1998)*

Region	Technically exploitable annual energy (TW-hr)	Annual hydro generation (TW-hr)	Energy exploited (%)
Asia	4330	367	8
Latin America	3540	367	10
Africa	1150	36	3
North America	970	536	55
Former USSR	3830	220	6
Europe	1070	485	45
Oceania	200	37	19
World total	15 090	2048	14

Adapted from: Moreira and Poole, (1993).
The percentage exploited generating power capacities do not correspond to the percentage exploited annual energy generation because of the wide variations in capacity factors of hydro plants from region to region.

thrust for development, would be a historic parallel to the industrialization of Europe and North America using fossil fuels. However, if environmental and social disruption are avoided in the process of industrialization, hydro development may reach a limit well below the technical potentials cited for the developing countries.

References

Anon. (1993). China breaks ground for world's largest dam. *New York Times*, 22 June, C1, C10.

Bourassa, R. (1985). *Power from the North – James Bay*, New York: Simon & Schuster.

Brown, L.R., Kane, H. & Ayes, E. (1993). *Vital Signs 1993 – The Trends that are Shaping Our Future*. Washington, DC: World Watch Institute.

Casey, R.C. (1993). Repair and upgrade of a 1950 vintage hydro plant. *IEEE Winter Meeting*, Paper 93-WM137-0.

Deudney, D. (1981). *Rivers of Energy: The Hydropower Potential* (World Watch Paper No 44). Washington, DC: World Watch Institute.

DOE (US Department of Energy) (1985). *Small Hydropower Development: The Process, Pitfalls and Experience*. Report DOE/ID/122 54–1. San Francisco, CA: Morrison-Kunudson Engineers.

DOE (1991). *Photovoltaics Program Plan, FY 1991–FY 1995*. Report DOE/CH10093–92. Golden, CO: National Renewable Energy Laboratory

Dowling, J. (1991). Hydroelectricity. In: *The Energy Sourcebook*, eds. Howes, R. & Fainberg, A. New York: American Institute of Physics.

EPRI (Electric Power Research Institute) (1982). *Transmission Line Reference Book*. Palo Alto, CA: Electric Power Research Institute.

Fischer, F. & Forester, J. (eds.) (1987). *Confronting Values in Policy Analysis*.

Beverly Hills, CA: Sage Publications.

Fischhoff, B. & Cox Jr, L.A. (1986). Conceptual framework for regulatory benefits assessment. In: *Benefits Assessment: The State of the Art*, Ch. 4. Dordrecht: Reidel.

Fritz, J.J. (1984). *Small and Mini Hydropower Systems – Resource Assessment and Project Feasibility*. New York: McGrawHill.

Globe & Mail (1991–92). James Bay Great Whale Project (daily newspaper, Toronto, Ontario: series running from April 1991 through February 1992).

Golof, R. & Brus, E. (1993). *The Almanac of Renewable Energy*. New York: Henry Holt.

Goodman, A.S. (1984). *Principles of Water Resources Planning*. Englewood Cliffs, NJ: Prentice-Hall.

Hingorani, N.G. (1988). Power electronics (In electric utilities: role of power electronics in future power systems). *Proceedings of the IEEE, 76,* 481–482.

Hock. S., Thresher, R. & Williams, T. (1992). The future of utility-scale wind power. In: *Advances in Solar Energy,* Vol. 7, Ch. 9, ed. Boer, K.W., pp. 309–371. Boulder, CO: American Solar Energy Society.

Hunter, L.C. (1979). *Water Power – A History of Industrial Power in the United States 1780–1930*, Vol. 1. Charlottesville, VA: University of Virginia Press.

IEEE (1990). Power and energy. *IEEE Spectrum*, Technology '90 issue, Jan., 44–45.

IEEE (1992). Tapping remote renewables. *IEEE Power Engineering Review*, 12, 3–28.

IWPD (International Water Power and Dams Construction) (1989). National Hydro Capacity. *International Water Power & Dams Construction*, Aug., 6.

IWPD (1990). The World's major dams and hydroplants. *International Water Power & Dams Construction*, Oct, 50–60.

IWPD (1991). The world's small hydro power. *International Water Power & Dams Construction*, May, 28–29.

Kern, E.C. (1993). Transmission worth of photovoltaic generation. *IEEE Power Engineering Review*, 12, 12–13.

Kozloff, K.E. & Dower, R.C. (1993). *A New Power Base – Renewable Energy Policies of the Nineties and Beyond*. Washington, DC: World Resources Institute.

Moreira, J.R. & Poole, A.G. (1993). Hydropower and its constraints. In: *Renewable Energy – Sources for Fuels and Electricity*, Ch. 2, eds. Johansson, T.B., Kelly, H., Reddy, A.K.N. & Williams, R.H. pp. 73–119. Washington, DC: Island Press.

Nozari, F. & Patel, H.S. (1988). Power electronics [in Electric utilities: HVDC power transmission systems]. *Proceedings of the IEEE*, 76, 495–506.

Park, P. (1992). Water tribunal rules on Cree homelands. *New Scientist*, Feb., 16.

Raphals, P. (1992). The hidden costs of Canada's cheap power. *New Scientist*, 15 Feb., 50–54.

Shaw, T.L. (1993). Environmental effects of hydropower schemes. *Proceedings of the IEEE, A*, 140, 20–23.

Thompson, M.S. (1980). *BenefitCost Analysis for Program Analysis*. Beverly Hills, CA: Sage Publications

Verhovek, S.H. (1992). Power struggle – a massive hydroelectric project could help light New York but may also obliterate a way of life in Quebec. *New York Times*, 12 Jan.

Williams, S. & Bateman B.G. (1995). *Power Plays – Profiles of America's Independent Renewable Electricity Developers.* Washington, DC: Investor Responsibility Research Center.

Wu, C.T. (ed.) (1990). AC–DC economics and alternatives – 1987. *IEEE Transactions on Power Delivery, Panel Session Report*, Vol. 5, 1956–1976.

Further reading

Anon. (1988). Extracting power from the Amazon Basin. *IEEE Spectrum*, Aug., 34–38.

Anon. (1997). A dam open, grand canyon roars again (Science Times Section). *New York Times*, 25 Feb., C1, C4.

Hammons, T.J., Blylan, B.K., Johnson, R. (1997). African electricity infrastructure, interconnections, and exchanges. *IEEE Power Engineering Review*, Jan., 6–16.

5 Energy storage

5.1 Introduction

Energy-storage technologies will play a critical role in the future adoption of sustainable energy-source technologies. Two of the most important renewable resources, solar and wind, are intermittent and until cost-effective means are found of increasing their availability over time, deep market penetration is unlikely. The consideration here is not solely technological. Storage capacity for an intermittent source requires investment beyond that necessary for exploitation of the source itself; consequently, it will affect the fixed costs of the energy delivered. In addition, if alternatives to petroleum-based transportation are to be found in EVs, then other means of automotive energy storage must be devised that are competitive technologically with the conventional fuel tank.

These two areas of application have been categorized as *stationary* and *mobile* storage technologies. The term stationary applies to immobile installations for electric power systems, industrial heat uses, and domestic (home) needs, whether in the conventional forms of the present or as the renewable energy sources of the future. Mobile storage applications might also be called "automotive" uses, denoting the means by which energy may be carried on board a moving vehicle to supply its own motive power, be it a fuel or electricity.

These two categories of storage applications have distinct technological requirements and it is well to define these before launching into assessments of specific technology prospects. In the stationary category, we are concerned in all cases with what has been called "secondary energy storage" (Jensen & Sorenson, 1984), which is storage needed in the process of energy transfer from the primary source to the end use. On this basis, *primary* energy storage would be the storage of the energy in the form of the primary resource (e.g. a fossil fuel or incident sunlight) prior to conversion to the energy form for end use. Electricity provides a clear example in that it has been generated from a primary source, such as solar, wind, or fuels, and must be transmitted instantaneously to its (end-use) load unless it can be stored in some way.

The best-known means of storing electricity are batteries or hydroelectric ("pumped") storage. Conversion of electricity to hydrogen, which we will define in Chapter 9 to be a secondary fuel, is another means under development. Strictly speaking, electricity *per se* is not stored by these means but rather is converted to another energy form (chemical or mechanical) to be stored and then converted back when needed again as electricity. The only devices that store energy in purely electrical (electromagnetic) form are capacitors or magnetic coils. Capacitors have never been developed for storing amounts of energy sufficient for practical use in power systems. They are being explored, however, to serve in a supporting role with batteries, as we discuss later briefly in this chapter. Magnetic coils, however, have prospects for power applications if superconducting coils are developed to the practical and economic stage, as we will also discuss later in this chapter.

Batteries and conversion to hydrogen will require further development if they are to play a significant role in support of the use of sustainable energy *sources*. Hydroelectric pumped storage, by comparison, is a working technology and used on conventional power systems. Its use with renewable sources is, therefore, an existing option that can be exercised wherever the investment appears justified, considering the large scale of conventional pumped-storage plants. Our focus here will be on the newly evolving storage technologies, such as batteries and hydrogen conversion, that could play a role in widening the possible applications of solar and wind sources, with installations large or small.

Heat provides the other common example of secondary storage when it has been produced from a primary heat source such as solar or fuel. Unless such heat is either transferred to its end use or stored it will ultimately be lost to its surroundings. Heat may be stored in water, steam, or other media for a limited time until, again, it is lost to the surroundings.

Energy can be stored not only in the forms of electricity and heat, but also as mechanical or chemical energy. The storage form appropriate for a given application will depend on both the source technology and the end use. For solar-thermal sources (Chapter 1) obviously heat storage is a possibility, while solar-PV cells would suggest electric (battery) storage. Energy can also be converted from one form to another for stationary storage; for example, in conventional pumped-hydro plants, electric energy is converted into the potential energy of a mass of water raised to a higher elevation (Cassedy & Grossman, 1990).

An important function of storage for the newly evolving stationary applications is to compensate for the intermittency of sources such as solar and wind. Solar sources, be they thermal collectors or (PV) electric, are subject to the regular diurnal and seasonal cycles onto which are imposed the random variations of weather. Wind-generator outputs also have diurnal and seasonal variations (more in some regions than others) and are dependent on weather sequences and fluctuations. Both have short-term fluctuations, on scales of seconds to minutes, such as passing clouds for solar collectors or gusting winds for windmills. Both can have longer-term variation, on the scale of days, such as cloudy days or windy days caused by a storm. In any of these cases, whether they are regular in time or random, energy storage can serve to smooth output or to insure the continuity of output from the renewable source.

The purposes of stationary storage for specific intermittent sources are varied (Winter, 1990). Short-time storage may be used to smooth out short-term operational fluctuations; this is sometimes termed buffering. Longer-time storage (over hours) can be used either to delay the period of

Figure 5.1
An electric utility
battery storage facility.
(Courtesy of Southern
California Edison.)

energy delivery ("period shift") or to extend the hours of delivery on a
given day. Delaying the delivery period is used to obtain a better match
with the times of use, whether for stand-alone service (e.g. domestic hot
water at remote sites) or for utility-grid service (e.g. solar electric feeding
an evening load on the grid). Extending the delivery period is a means of
increasing the solar fraction of solar-thermal systems operating with fuel-
burning back-ups.

Electric-energy storage is already in use in conventional power systems for "load-leveling" operation. Figure 5.1 is an example of a prototype battery installation operating on a utility grid system for load leveling. Load leveling attempts to smooth generator output requirements over the daily variations in load demand. The objective, historically, has been to minimize the sum of operating and capital costs (Stoll, 1989) of conventional fossil and nuclear plants. One might expect, at first glance, that the addition of renewable sources (wind or solar) to otherwise conventional power systems might demand additional storage capacity on the system to maintain system reliability and to make the most economic use of conventional generation. This notion has been challenged, however, for power systems that have only partial penetration of intermittent renewables (Kelly & Weinberg, 1993). A clear-cut example of this is the addition of solar generation in the sunbelt region where peaks of solar input coincide with peaks of air-conditioning demand on the electric grid system. Our concern here, however, will be less on the system aspect of storage and more on the (secondary) source aspects and on the economics thereof.

The mobile storage applications are for the electric vehicle, supplied by either batteries or fuel cells that are run off (stored) hydrogen fuel. Here, the technological requirement is for stored energy that is portable in sufficient quantities to meet the speed and driving range specifications normally expected for cars and trucks. Portability is determined by the density of the storage, in terms of *specific power* and *specific energy*, which are measured in units of watts/kilogram and watt-hours/kilogram, respectively. Specific power will determine the speed (or acceleration) of an electric vehicle, whereas specific energy will determine its driving range.

Our focus here will be on the storage technologies themselves: their technological and economic prospects for enhancing the renewable source technologies. As such, we will be concerned with the operating characteristics and costs in storing and delivering a unit of energy for each storage form. Most of the prospective stationary, storage technologies we will consider, be they electric or thermal storage, are technically operable. Most at present, however, are unfeasible economically. By comparison, mobile storage must be considered as marginal technologically because of the poor specific power/energy characteristics of the prospective forms of batteries or the mobile storage of hydrogen for fuel cells. Environmental policy pressures for alternatives to the conventional ICE vehicle place a priority on technological performance for electric vehicles, rather than on economic competitiveness. Consequently, as discussed in Chapter 2 with biomass fuels, if working technological alternatives can be found to the gasoline (or diesel) engine for transportation, then assessment of cost competitiveness may have to be adjusted to encompass public policy measures.

5.2   Batteries: electrochemical storage

Dramatic developments are needed in electrochemical technology to create batteries that are much lighter, much less bulky, and much longer lived in order to capture markets for the electric energy-storage applications discussed in our introduction. The design and production methods must also create a much less expensive product. These deficiencies are a long-standing challenge for the electrochemical industry.

Batteries are the oldest electrical technology, dating back to the cells of Galvani and Volta at the end of the 18th century (Jensen & Sorenson, 1984). In modern terms, batteries are electrochemical systems in which electrical energy can be converted into the energy of a chemical reaction and then be retrieved through a reverse reaction. In these reversible reactions, the chemical composition of the electrodes changes during discharge and then changes back in the recharging process, with electrons and ions migrating through an electrolyte to complete the process in each direction.

The common lead–acid battery uses electrodes of lead and lead oxide, with an aqueous sulfuric–acid electrolyte (Fig. 5.2a). Lead–acid batteries are the conventional secondary (rechargeable) battery technology today and are used for starting automobiles and for a variety of remote stationary uses. In most usage, batteries are not required to store large amounts of electric energy. The majority of non-utility applications require a fraction of a kilowatt-hour storage capacity, with only some heavy-duty requirements involving several kilowatt-hours. The largest lead–acid battery installations to date have been electric utility test projects in the 10–50 MW-hr range, which have been conducted in anticipation of later advanced-battery developments for load-leveling application (Anderson, 1993).

Lead–acid batteries are unsatisfactory for stationary use with intermit-

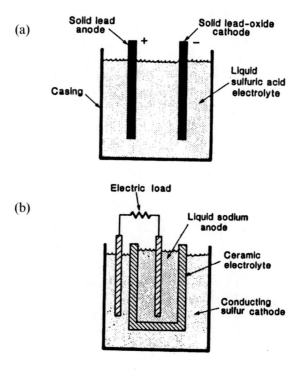

Figure 5.2
Secondary (rechargeable) batteries. (a) Lead–acid battery. (b) Sodium–sulfur battery. (*Source*: Cassedy & Grossman, 1990.)

tent sources because of their short lifetimes, large bulk, and high cost. They are unsatisfactory for mobile (automotive) use because of their short lives and the weight and cost penalties they bring to electric vehicles (McHarnon & Cavins, 1989). Short lifetimes are manifested in the limited number of "deep-discharge" cycles that the battery can undergo before the electrodes deteriorate. Unlike the service required of automobile batteries, which are rarely discharged over half their capacity, energy-supply applications require repeated discharge of most (e.g. 80%) of their capacity. Such service, whether for renewable sources or for electric vehicle use, would typically be on a daily basis and therefore, 250–350 deep discharges could be expected annually.

For many years, lead–acid battery lifetimes have been less than 1000 deep cycles, meaning a life span of 3–4 years before replacement. Recent developments suggest a possible 2000 deep cycles (McHarnon & Cavins, 1989; Chan, 1993) implying a life span of 6–8 years. This results from designs that reduce electrode corrosion and improve the circulation of the electrolyte. Another battery type in commercial use in Europe, nickel–cadmium, has a life-cycle of about 2000. It also has a specific energy 30% greater than lead–acid but is several times more costly (IEA, 1993).

Even with lighter-weight electrodes in the advanced lead–acid batteries, their excessive mass still limits the automotive performance of EVs (Chan, 1993). The weight-specific characteristics of EVs using lead–acid storage must be compared with conventional automobiles that carry their stored energy as fuel in a tank. Conventional automobiles and light trucks have specific energies in the hundreds of watt-hours/kilogram and specific powers over 500 W/kg, resulting in driving ranges over 350 miles (560 km) and maximum speeds in excess of 120 mph (196 km/h). EVs using lead–acid batteries have reached specific energies of only about a fifth and specific powers less than half those of conventional automobiles. This results in driving ranges for EVs of only a little over a third of that of gasoline-fueled automobiles and maximum speeds of about half.

Whereas, this higher automotive performance might not be considered essential to public tastes more attuned to sustainable life styles, the gap in technology is not likely to make electric vehicles widely acceptable in the absence of a change in tastes (Brown, 1995). Technological improvements are, therefore, being sought involving other electrochemical combinations such as sodium–sulfur, lithium–iron sulfide, zinc–bromine, nickel–metal hydride or metal–air. These are being investigated in R&D programs in the USA, Europe, and Japan (Chan, 1993; MacKenzie, 1994; IEA, 1993). Some of them have been under investigation for decades, however, without dramatic results.

Sodium–sulfur batteries (Fig. 5.2b), by not using solid electrodes that can decompose, are expected to show better life-cycle performance. In addition, by using elements lighter than lead, the specific energy/power figures can be improved. Indeed, prototype sodium–sulfur cells display 100–165 W-hr/kg (45–75 W-hr/lb) and 130–220 W/kg (60–100 watts/lb), which exceed the figures for lead–acid batteries by as much as 4:1 in specific energy and up to 2:1 in power. So far, however, sodium–sulfur cells have not achieved longer life-cycle times than advanced lead–acid cells.

Even if sodium–sulfur batteries were to approach competitive performance, however, they have the disadvantage of high-temperature operation. The use of a molten sodium electrode and a molten beta-alumina electrolyte requires operation at 300–350 °C (572–662 °F). The presence of this very hot molten material and the possibility of violent chemical reactions of the sodium with water pose hazards for use of these cells in automotive service. Another possibility, the lithium–iron sulfide cell, also requires a molten electrolyte at a temperature of 450 °C (842 °F) and, therefore, poses the same hazards. In addition, this cell does not perform as well in its life cycle and does not have the specific weight parameters of the sodium–sulfur cell (McHarnon & Cavins, 1989).

A sodium–nickel chloride combination is another possibility for a high-temperature battery (MacKenzie, 1994). Its specific powers and energies to date are comparable to sodium–sulfur cells and accident hazards are less because the active materials are all solid. In addition, it has promising life-cycle features and less-stringent preheating requirements. The costs of this combination are high compared with other advanced battery possibilities.

The zinc–bromine and metal–air cells, by comparison, operate at ambient temperatures and offer prospects for improvements in performance. Zinc–bromine cells have specific energies nearly twice those of lead–acid and comparable efficiencies (75–80%) (Chan, 1993; IEA, 1993). They have, however, a relatively low peak-power density (53 W/kg), a short shelf life, and a potential hazard of leakage of noxious bromine gas. The metal–air cells create free oxygen upon charging and react with the oxygen in the air upon discharge. Also, they use iron, zinc, or aluminum electrodes that are replaceable after corrosion sets in. Both of these cells need further technical development as they currently suffer from either low efficiency or material instability.

Nickel–metal hydride batteries promise good technical performance for EV use (Chan, 1993). They have a power density over 200 W/kg (91 W/lb), which is exceeded only by the high-temperature batteries, and have a relatively long life cycle (over 1000 cycles). In addition, they can be recharged more rapidly than other combinations and require little or no maintenance. The present costs of these batteries are, however, nearly 20 times that of lead–acid cells and are greater than any of the other advanced batteries. Recently, electric vehicles using this battery type have turned in the best driving range performances over standardized race courses (NESEA, 1997) (more on this below).

New devices called ultra-capacitors are being developed that can provide an improved power delivery to enhance the acceleration performance of EVs (Burke et al., 1990; Dowgiallo & Harding. 1995). These devices can be considered as transitional in construction (Conway, 1991) between simple capacitors, where charge is collected on two opposing electrodes, and batteries, with conversion of the electricity by chemical reactions. In their present stage of development, however, ultra-capacitors can only be used to supplement batteries with short bursts of power to aid the acceleration of the EV. They cannot store enough energy to provide any driving range to the vehicle. It is possible, however, that this research could progress to re-design of EV batteries with more emphasis on energy storage and less on power delivery, on the assumption that the capacitors could serve the power-delivery needs. Even if this proved feasible technically, however, the added cost of the capacitor could be a barrier to adoption.

Other electrochemical combinations have been investigated (McHarnon & Cavins, 1989; Chan, 1993), but none of them has achieved the required performance in life cycles or weight-specific parameters. Some of them have disqualifying characteristics of highly toxic components (e.g. cadmium in nickel–cadmiun cells). In addition, parameters of efficiency and maintenance requirements must be considered for any candidate technology. Efficiency refers simply to the round-trip (charge–discharge) ratio of energy out per unit of energy in, for which most applications would require 60% or better. Lead–acid batteries will deliver as much as 80% of their charged energy input. Table 5.1 shows the relative gains in performance, for EV service, of several batteries under development compared with lead–acid batteries.

Returning to batteries for stationary uses, the considerations are more limited than for mobile uses. Here, we are concerned mainly with life cycles and costs and secondarily with bulk. If batteries could be developed that were cyclable thousands of times on deep discharges, at low cost, then occupying large volumes to do so would not be disabling for most applications. Therefore, it *is* of prime importance that the life cycles of batteries that are currently available commercially are less than 2000 (deep-discharge) cycles and that storage adds significantly (even prohibitively in some cases) to the cost of energy delivered in stationary applications.

5.3   The costs of electric energy storage

The costs of energy storage can be considered from various perspectives, depending on the application. The criteria for systems applications, such as utility load leveling, should cover the generation costs of the various sources making up the generation mix (Stoll, 1989) of the grid at hand. When considering storage of energy supplied from a single (renewable, intermittent) source, however, cost criteria can be confined to the energy supplied by that source alone. The most clear-cut example is stand-alone (solar or wind) generation not connected to a utility grid, where the cost consideration is purely the cost of a unit of energy (kilowatt-hour) as delivered from storage. The total cost of such stand-alone energy must then be the sum of its conversion (or generation) cost *plus* its storage cost. If this total cost is high, this is a barrier to adoption of the source–storage combination.

"Period displacement" (Swet, 1987) in utility operation is conceptually near to the stand-alone case in simplicity and provides another illustration of storage costs added to generation costs. Period displacement refers to storing energy during the time period it is available (e.g. solar energy at

Table 5.1. *Comparisons of developmental batteries with lead–acid batteries*

	Battery type						
	Pb–acid	Ni–Fe	Ni–Cd	Ni-metal hydride	Na–S	Li–FeS$_2$	Li-polymer
Relative energy density	1.0	1.5	1.6	1.7	2.5	4.0	4.0
Relative peak power density	1.0	1.2	1.9	2.1	1.1	4.0	3.5
Relative range	1.0	2.0	2.1	2.3	3.4	4.0	4.0
Energy efficiency (%)	68	58	80	76	91	80	85

Source: adapted from Chan (1993).

midday) and then delivering it later when the demand occurs (e.g. evening household needs). It is useful in utility grid-connected operation where the load demand on the grid is not concurrent with (solar or wind) generation. Here too, the unit cost of energy delivered (to the grid) must be the total cost: generation plus storage. This cost, as delivered, will be the determinant of whether an intermittent source (with storage) fits economically into the generation mix of the grid at a particular point on the time profile of the utility's load.

In the operation of either PV or wind generation with utility grids, electricity storage can be considered as a way to increase the operating availability of these intermittent sources (Kelly & Weinberg, 1993). To the extent that storage increases the short-time availability of PV or wind generation on days when the (sun or wind) resource itself is available, operational cost reductions can be achieved by lowering the "spinning-reserve" back-up normally required for reliability of the grid. In addition, increased availability of the renewables will enhance the displacement of high-fuel-cost generation for peak loads. This is likely to be cost effective not only through these operational savings but also by achieving a higher capacity factor of utilization on the investments for storage.

Going to higher storage capacities to obtain longer-time energy carry overs, however, is likely to have diminishing returns. Under the present state of electric-storage technology and costs, economic operations seem to be restricted to short-time storage functions. Consequently, until there is a dramatic drop in the costs of energy storage, the percentage penetration of the market by intermittent sources with small or zero storage will be limited. According to most system studies, penetration will not get beyond approximately 15% (Chowdhury, 1988, 1991). There are other studies, however, that show more optimistic projections, in some instances going to 30%, for example for a new utility system with an optimum generation mix (Grubb & Meyer, 1993; Kelly & Weinberg, 1993). The most optimistic case study suggests an upper limit of 45% of electricity supply (energy delivered, not capacity credit) on particular wind-farm installations without the benefit of storage.

Now, taking the economic assessment of storage to more quantitative comparisons, consider the case of photovoltaic electricity or wind generation with battery storage, either in stand-alone or period-displaced grid operation. Recently, available lead–acid battery systems have had capital costs (per unit of energy–storage capacity) in the range \$150–200/kW-hr (Culta, 1989; Humphreys & Brown, 1990; Ingersoll, 1991). These figures include both power-related and energy-related costs of the storage system, where the energy-related costs include additional expenses in battery replacement. Advanced lead–acid batteries are currently assumed to have life cycles ranging around 1500 deep discharges and would, therefore, have to be replaced after 6 years assuming 250 discharge cycles per year, or in as little as 4 years if daily discharges are required. Using these figures with a 10% discount rate, the levelized cost of electricity delivered from storage on each discharge cycle (with a 66% system efficiency) is in excess of \$0.20/kW-hr, *not* counting the cost of the PV or wind electricity. This clearly prices battery storage out of the market for most applications at the present time.

Looking now to the future, advanced lead–acid batteries are forecast (Kelly & Weinberg, 1993) in the medium term to have life cycles up to 3750 deep discharges, meaning replacement in 10–15 years, depending again on the annual discharge cycles. This extension of life cycle alone, if it comes to pass, would result in a dramatic drop in levelized cost (again, for energy delivered) to figures approaching $0.08/kW-hr (Humphreys & Brown, 1990). Even with such a reduction in storage costs, however, the *total* cost of PV electricity, as delivered from storage, would be in the range $0.20–0.28/kW-hr for the midterm projected cost of PV conversion of $0.12–0.20/kW-hr in the year 2000 (see p. 55). Such a cost of delivered electricity is, of course, substantially out of the likely competitive range for bulk-power applications, leading us to conclude that *battery storage is unlikely to be a major aid to the penetration of photovoltaics into the grid electricity market in the midterm future.*

The use of batteries with wind generation may tell a different story, however, since generation costs there are already considerably below PV costs. Wind-generated electricity has a levelized cost that has fallen below $0.09/kW-hr (Chapter 3), which makes for a total cost of $0.29/kW-hr as delivered from storage in presently available lead–acid batteries. While this appears not to be competitive in the near term, the projected medium-term wind-generation costs of around $0.05/kW-hr make wind and storage a near-competitive combination. This is seen when we combine this projected generation cost with the medium-term forecast $0.08/kW-hr for lead–acid batteries to make a total cost of $0.13/kW-hr. Therefore, it would appear that projected battery-storage technology would *likely be an aid in the further penetration of wind generation into both grid and stand-alone markets in the medium-term future.*

Major innovations are needed in battery storage before it can be an aid to market penetration of PVs for further stand-alone applications or into significant grid operations. Longer-term forecasts for other advanced batteries under development hold some promise for cost reduction (McHarnon & Cavins, 1989), albeit uncertain. Sodium–sulfur batteries may attain lifetimes of 5000 cycles at 75% efficiency, with initial costs of around $100/kW-hr. If attained, this could mean a stored energy cost for stationary applications of less than $0.05/kW-hr, assuming the same financing basis as above and no major fixed costs (e.g. replacements) over a battery lifetime (15–20 years). Such a storage-delivered cost, together with long-term PV goals (DOE, 1991) of $0.05–0.06/kW-hr, could conceivably result in the competitive $0.10–0.11/kW-hr for total cost of PV electricity as delivered from storage. Again, however, it should be remembered that such forecasts for PVs have been shown to be overly optimistic, especially with regard to the ability to meet projection targets on time (see p. 56).

For mobile applications (EVs), most of the advanced battery prospects have higher initial costs at present than lead–acid batteries, some of them an order of magnitude higher. Projections of costs for EV batteries for the year 2000 show only the sodium–sulfur battery reaching down to the

Table 5.2. *Criteria for advanced electric vehicle batteries produced by the US Advanced Battery Consortium*

Criteria	Mid term	Long term
Power density (W/l)	250	600
Specific power, (W/kg)	150	400
Energy density (Wh/l)	135	300
Specific energy, (Wh/kg)	80	200
Cycle life (cycles)	600 (5 years)	1000 (10 years)
Unit price (US$/kW-hr)	<150	<100
Operating environment (°C)	−30 to 65	−40 to 85
Recharge time (h)	<6	3 to 6
Efficiency (%)	75	80
Self-discharge time	<15% in 48 h	<15% per month
Maintenance	Zero	Zero

Source: adapted from Chan (1993).

same cost level as lead–acid cells and achieving the ultimate price levels necessary to fit the criteria for adoption recognized by the auto industry and government. Criteria for the required cost, life, and maintenance, as well as for performance, for advanced EV batteries have been listed by the US Advanced Battery Consortium (CEC, 1991; EPRI, 1991; Barber & Abarcar, 1992; Chan, 1993). The Consortium was founded in 1991 as a collaborative government/auto industry effort to further battery R&D. These criteria are shown in Table 5.2. In addition to the unit costs shown, the Consortium has set cost criteria per battery of $6000 by 1995 and $4000 by 2002. It should be kept in mind that these criteria apply to electric vehicle service, which has particular requirements of ruggedness, safety, re-charge time, and maintenance. In addition, these cells are cycled in a different pattern to those in stationary service. The best measure of the battery lifetime in automotive service is not the total number of deep cycles but rather the total vehicle mileage it will supply for the vehicle over the life of the battery. On this basis, an EV battery deemed acceptable by the Consortium for the medium term is one that supplies 50 000 miles (80 000 km) of vehicle travel, even though it must be replaced after 750 deep discharge cycles rather than the 2000–3000 targeted for stationary applications.

Responding to the policy demands for low/zero-emission vehicles (CEC, 1991; Ford, 1994; Wald, 1994), the automotive industry has produced prototype electric vehicles using the battery technologies available. A comparison of the claims for performance of five of these models is given in Table 5.3. Some of these performance figures are the ratings of the manufacturers (Chan, 1993), while others include the recorded performance at the 1997 Tour de Sol (NESEA, 1997). The critical figure for general use, of course, is the range in comparison with cars run by ICEs. It should be

Table 5.3. *Battery and performance comparisons of the latest electric vehicles*

	FEV	BMW El	GM Impact	Ford Ecostar	Solectria Force
Battery type	Ni–Cd	Na–S	Pb–acid	Na–S	Ni–metal hydride
Maximum speed: mph (km/h)	80(128)	75(121)	75(121)	70(113)	75(121)
Range: miles(km)	155(248) (25 mph, 40 km/h)	135(217) (55 mph, 88 km/h)	120(193)	100(160) (rated); 200(320) (Tour de Sol)[a]	150(240) (rated); 249(400) (Tour de Sol)[a]
Hazards	Toxic	High temperature		High temperature	

[a] At speeds determined by the Tour de Sol course.
Source: adapted from Chan (1993) and NESEA (1997).

noted that the longest range, of the three manufacturer's claims, has been done at such a low speed (25 mph) that the average car buyer would consider it impractical. The Tour de Sol range results were derived from a specified tour course that took the vehicles over varied town and country (not interstate) highways and, therefore, represents more average driving conditions. It should be noted, however, that other conditions for these vehicles were not entirely comparable with conventional automobiles, such as the load they carried over the tour route. The "Selectria Force" vehicle has a load capacity of only 450 lb about half that of a conventional sedan. It achieved the record range of 249 miles with a total curb weight of 2500 lb (1136 kg), about three quarters that of a standard sedan with an ICE. The rated range of this EV, as it is produced for the market, is 150 miles (240 km) at the upper end of claims by other EV manufacturers.

Another indication of practical performance comes out of the Clean Fleet prototype operations in California (BMI, 1995), where lead–acid and nickel–cadmium batteries were used in fleet operations for delivery and service tasks over the period 1992–94. These EV vans carried battery packs weighing 850–1144 kg (1874–2522 lb) and had ranges of about 25 miles (40 km) for the lead–acid batteries and 50 miles (80 km) for the nickel–cadmium batteries. Although these performances were also influenced by the electric-propulsion technology, they nonetheless depended fundamentally on the specific energy/power parameters of the vehicle, which, in turn, were dominated by the weight of the batteries. Finally, it should be noted that none of these prototype EVs is near an affordable cost to the average consumer, principally because of the cost of batteries and their replacements over the lifetime of the vehicle.

The best performing EVs for range in both the 1996 and 1997 Tour de Sol races were powered by nickel–metal hydride batteries and the second best in these 2 years used sodium-sulfur cells. In 1998, the best performing EV was again powered by a nickel–metal hydride battery (NESEA, 1998). Not withstanding the disadvantages of toxic hazards of some of these advanced batteries, it is clear that progress has been made in this technology toward the technical requirements for EVs (battery). However, this progress has been achieved without regard to cost, since these batteries currently typically have initial unit costs (at $1500/kW-hr) an order of magnitude higher than lead–acid batteries. In order to compare EV costs or performance for specific tasks, the automotive service or "mission" must be defined (Humphreys & Brown, 1990). The service categories for these comparisons for EVs must all be for light vehicles operating over the limited driving ranges that the present state of the technology restricts them to. The passenger-car mission, for example, provides for a payload of only 300 lb (140 kg) and a driving range of 250 miles (400 km), while the mission for a delivery van requires a 650 lb (300 kg) payload and a 60 mile (100 km) range. It should be recognized that delivery missions are restricted typically to those for light delivery in urban commercial use.

Passenger uses of these EVs can only be of limited-range suburban or urban driving. Even an urban taxicab service, which averages 250 miles (400 km) per day (NRC, 1976), could not be met with the present technology. It has been estimated that UPS (a parcel delivery service in the USA) would find that only about one quarter of its routes would be serviceable by EVs with the present technology. Long-distance service, such as intercity, passenger-car travel, or freight hauling is, of course, out of the question without major improvements in battery energy densities and recharge times. Nonetheless, steps have been taken by the electrical and auto industries, with government help, to anticipate the need for an EV-charging infrastructure, at least in urban areas where air pollution measures will be mandated (Moore, 1993). Prototype charging stations have been set up in Los Angeles and other urban areas where the density of users would justify their operation.

Finally, for battery-powered electric vehicles, it is found that life-cycle costs (for the entire vehicle, with storage) are presently higher than those of ICE vehicles, when measured on a per mile basis. While this measure does not account for the present differences in performance of EVs and ICE vehicles, it nonetheless gives a useful comparison for comparable missions. For the light-duty missions, EVs using conventional lead–acid batteries have levelized, life-cycle costs estimated at $0.051/mile ($0.032/km), compared to a nominal $0.043/mile ($0.027/km) for ICE vehicles (Humphreys & Brown, 1990). A more advanced "sealed" lead–acid battery model has an estimated cost ($0.042/mile, $0.026/km) slightly less than the ICE car. No better is a sodium-sulfur, battery-powered van; although this has a superior EV life of 1000 cycles, it nonetheless has a life-cycle cost of $0.063/mile ($0.039/km). Consequently, it can be seen that costs for EVs are in a range comparable but somewhat higher than conventional ICE vehicles, but only for the light-duty, limited-range missions.

5.4 Flywheels: electromechanical storage

Developments in recent years appear to be moving the flywheel from an interesting phenomenon toward practical use as a means to store energy. The developments have followed advances in high-tensile materials and power electronics (Post *et al.*, 1993), and could lead to a competitor to the electrochemical battery. The most promising sustainable technology prospects appear to be for mobile storage in electric vehicles.

Energy can be stored in flywheels in the form of the kinetic energy of a rotating mass (Ingersoll, 1991). Historically, flywheels were constructed as massive rings of steel or iron and were first used to smooth the rotation of reciprocating steam engines. Such smoothing action, also found in rotary automobile engines, does not store a significant amount of energy relative to the power required for locomotion. Attempts to store energy in mobile form for automotive use have been made in recent years in Europe and the USA (Post, *et al.*, 1993). The outcomes of this research, until recent years, had yielded specific energies (kilowatt-hours per kilogram) too low to be practical.

Research efforts in the 1970s, sponsored by the US Government, laid the ground work for significant progress a decade later. The most successful flywheel rotors have been constructed of concentric cylinders made of high-tensile-strength fibers embedded in epoxy resins. The high-tensile-strength fibers permit ultrahigh rotational speeds without the need for high-mass rotors. Using these structures, specific energies over 500 W-hr/kg have been achieved, thus exceeding those of any of the advanced electrochemical batteries by a ratio of over three to one. Some results suggest that this ratio will go to four to one. The key to these advances came through the development of new superstrong fibers, that allowed operation of flywheels at extremely high rotational speeds (e.g. 200 000 r.p.m.). Since kinetic energy in rotating masses increases as the *square* of the rotational velocity, disproportionally high increases in stored energy are to be had for each increase in rotational speed (Genk, 1985). Thus, the new route to increasing stored energy was to raise the speed rather than the mass. The faster the rotation, however, the greater the stresses within the flywheel from centrifugal forces. Earlier research efforts had resulted in catastrophic shattering of flywheels from rotational stresses. The development of the stronger fiber-based materials allowed operation at the high speeds.

These high-speed flywheels also have operational advantages over batteries, including higher efficiencies and higher power rates of energy transfer (both on input and output). The high efficiencies are achieved by the use of virtually frictionless magnetic bearings and nearly-zero windage inside an evacuated housing. Operation with losses of 1% or less is thought to be possible, compared with 20–25% losses for electrochemical batteries.

An ingenious electric motor/generator is used to couple electric energy into and out of the flywheel. It uses permanent magnets with fields coupled to stator windings through the vacuum envelope, as the magnets are rotated with the flywheel. The magnetic coupling is similar to that of a conventional a.c. machine, having three phases but of variable frequency. The three phases can be synchronized to conventional (fixed 60 Hz frequency) three-phase lines by the use of solid-state switching transistors. This use of "power electronics" is but one of several examples of solid-state devices that can facilitate the utilization of alternative-energy technologies.

The possibility also exists of returning energy to the flywheel, that is normally lost in braking, by means of "regenerative braking". Regenerative braking has the vehicle's wheels drive the electric machine as a generator, thus returning some motive energy to the flywheel. Such a scheme has already been proposed to improve the efficiencies of subway trains, where braking returns electric energy to the central power supply of the transit system. In flywheel vehicles, the return would be of mechanical energy to the flywheel on board. An alternative use would be to have the braking drive a generator (as it would for subway trains) and thus return electric charge to EV batteries (Jefferson, 1993). Regenerative braking could, thus, have a major benefit for automotive flywheels or EVs and thus increase the vehicle range for either means of on-board storage.

Flywheels, recently dubbed "electromechanical batteries" (EMBs) (Post *et al.*, 1993), offer a challenge to electrochemical batteries for any of the applications we have considered, whether mobile or stationary. Further development is required on technical issues such as vibrational stability and failure modes. Extensive field tests would be required to establish the working feasibility of any of the possible applications. Included must be safety studies, since stress failure of the wheels is still a possibility, even with the new fiber construction. Also, the costs of flywheels are still largely unknown. It is known that the special fibers, such as graphite fibers (the most successful) are expensive to produce. Nonetheless, this prospective technology offers an example of the unexpected outcomes that can come out of disparate research approaches.

5.5 Hydrogen storage

Hydrogen is said by its promoters to be the ideal storage medium for either stationary or mobile applications (IAHE, 1990). For stationary applications, it must serve as a secondary energy storage, as mentioned earlier. The primary source of energy in such cases would be a fuel (e.g. natural gas) or a renewable source (e.g. solar energy), as discussed in Chapter 9. For mobile applications, it can provide the portable means of storage to supply a vehicle powered either by an ICE using hydrogen as the combustible fuel or, more likely, a fuel-cell EV drive.

For the stationary applications, our prime interest here is in hydrogen storage that will serve in conjunction with renewable source technologies. The solar–hydrogen technologies discussed in Chapter 9 are clear-cut examples, where the primary sources of energy are solar PV cells, biomass fuels, and the like. In the most grandiose conception – the "hydrogen economy" (Gregory, 1973; Dickenson *et al.*, 1977) – hydrogen would be piped around the same way that natural gas is now, with massive storage depots to assure the continuity of supply to the pipeline network. In more modest schemes, however, hydrogen may serve the same technical function as batteries in providing either smoothing continuity to stand-alone inter-mittent sources (e.g. solar or wind) or a delay of the period of energy delivery in grid service. While only a few prototype projects have been carried out using such buffering or period-shifting functions for hydrogen storage (Ogden & Nitsch, 1993), this is a technological alternative to batteries for stationary storage applications. If such operation were to take place, the cost of energy delivered would again have to be the sum of source plus storage costs, as it was for batteries. For electrical applications, however, hydrogen storage requires additional steps and investments that batteries do not require.

The use of hydrogen storage requires first the conversion of the source electrical energy into hydrogen, for example by electrolysis. If the energy is to be delivered out of storage to supply an electric load, then the hydrogen must be converted back to electricity. The conversion of hydrogen to electricity is accomplished using fuel cells and the cost of the delivered electricity must then also include the (levelized) costs of the fuel cell. (The operation of fuel cells is discussed briefly in Chapter 9.) In view of these extra steps of conversion and reconversion, it is not surprising that a much higher total cost of delivered electricity results. This cost has been projec-ted to be nearly three times that of using battery storage with PV sources (DOE, 1992a).

The added cost owing to reconversion could only be avoided if the hydrogen output is the fuel for combustion only. This situation would fit the scenario of the solar–hydrogen fuel, as promoted by the hydrogen advocates (Chapter 9), but only in situations where the hydrogen would not have to face market prices for a gaseous fuel. This would obtain to schemes using solar PV as the stand-alone source for electrolysis and also for stand-alone wind–hydrogen, either of which could tolerate higher unit costs for the hydrogen output. In any stationary application using the output hydrogen fuel for combustion alone, once the electricity has created the hydrogen, the costs of hydrogen storage would be in relatively small increments over the production costs of the hydrogen (Ogden & Nitsch, 1993). The additional costs are only around $1/MBTU ($1/GJ) for storage compared with a production cost of approximately $23/MBTU ($24/GJ) for the hydrogen (Chapter 9).

When we turn to mobile applications for hydrogen storage, we are perforce talking about the fuel-cell vehicle (FCV), if we omit ICE use of the hydrogen. (ICE combustion of hydrogen in air would emit nitrogen oxides and, therefore, probably be excluded from the category of low-emission vehicles.) The FCV is a competitor to the battery EV. For the fuel-cell technology, as for batteries, prime consideration must be given to the technological feasibility of achieving a working alternative to conventional ICE vehicles. A promising fuel-cell candidate is the photon-exchange-membrane (PEM) cell, which has an advantage of a lower operating temperature than other fuel cell types (Miller, 1996). Here we will be concerned with the working features of the *mobile* storage of hydrogen and with the fuel cell.

In a FCV, the fuel cell supplies electricity (d.c.) to the traction motors driving the wheels, just as the battery does in an EV. While having much the appearance of a battery, the fuel cell does not have the energy stored within in a chemical form as does the battery. Rather, the cell must be fed a fuel (hydrogen) that enters an electrochemical reaction evolving electricity at its electrodes (see Fig. 9.3 and the discussion of the fuel cell in Chapter 9 for further details).

At present, FCVs have performance limitations, similar to battery EVs, relating to portable energy and power per unit of mass or volume, that limit their driving range and maximum speed. The mass determining the specific energy and power of FCVs depends on the weight of the means of storing the hydrogen on board. The storage possibilities are as a pressurized gas, as a cryogenic liquid, in metal hydride compounds, or adsorbed onto the surfaces of carbon particles (Jenson & Sorenson, 1984). Of these, the last two are still experimental.

Hydrogen gas can be safely compressed at very high pressures (thousands of pounds per square inch) into portable tanks suitable for automotive use. The energy volume density of the pressurized hydrogen, however, is an order of magnitude lower than that of gasoline. This means that the hydrogen tank would have to be nearly ten times the volume of the gasoline tank to carry the energy required for the same driving range. Actually, the higher conversion efficiency of the fuel cell (Kordesch & Oliveira, 1988) might reduce the required volume to only about five times that of gasoline.

The more advanced methods of hydrogen storage will require on-board energy inputs for their operation if they are finally developed for practical mobile use (McHarnon & Cavins, 1989; Ogden & Nitsch, 1993). Hydrogen can only be liquefied at cryogenic temperatures (approximately −250 °C) and, therefore, would require constant energy input for refrigeration on board. Carbon adsorption also requires low-temperature (−158 °C) refrigeration. (Adsorption is the process of chemical reaction of a gas with a solid at its surface, similar to that used in catalytic converters for automobile exhausts.) Finally, metal hydrides require heat to release hydrogen which has been absorbed into the hydride compound when stored (absorption of hydrogen into hydrides, such as zirconium–manganese and iron–titanium involves a deep penetration of the hydrogen atoms into the interstices of the solid lattice of the metal). These energy requirements all add to operational complexity and exact an energy penalty from the on-board system, in addition to increasing the initial capital cost of the system.

These storage technologies for hydrogen all have higher energy densities than batteries but all still are less than the conventional ICE vehicle

(Jenson & Sorenson, 1984; DOE, 1992a). For easy comparison, an average conventional ICE vehicle can travel about 355 miles (570 km) with a (gasoline) storage system (fuel plus tank) weighing 100 lb (45 kg). For a compressed-gaseous hydrogen storage system of 100 lb, a FCV (of current technology) can travel only 36 miles (58 km) and a 100 lb lead–acid battery would carry an EV only 5 miles (8 km). Advanced technologies have been projected (DOE, 1992a) to increase the gaseous hydrogen FCV's range to 275 miles (442 km) for this same 100 lb fuel load, but this is still marginally acceptable for performance. Hydrogen storage by cryogenic liquid or by carbon adsorption offer prospects for improvements in the energy densities of hydrogen storage, but the costs of such systems cannot be determined until R&D brings costs closer to commercialization.

When commercialization is considered for any of the mobile storage technologies, be they batteries or hydrogen/fuel-cell systems, critical choices will also have to be made with regard to practical requirements, such as fill up and start up. Again, the comparisons will be inevitable with the conventional gasoline-fueled ICE vehicle, which can be completely re-fueled in 2–3 minutes and started in a matter of seconds. Battery-driven EVs will require 6–8 hours to fully recharge from deep discharge, regardless of battery technology, if a "slow-charge" rate appropriate to existing residential wiring is used. Quick charges (30 minutes or less) would require special electrical service installations capable of supplying 60 kW power at a minimum per vehicle (i.e. 500 A at 120 or 250 A at 240), to supply an EV with a full supply of 30 kW-hr of energy (Chan, 1993). The recharging scheme most often advocated for private vehicles is slow charging at the owner's residence overnight during the off-peak times of electric utilities. This, of course, limits the daily driving of the electric vehicle to the driving range of that vehicle with a single, full charge.

Gaseous hydrogen fuel tanks can be refilled under pressure in about the same time (2–3 minutes) as conventional gasoline tanks. It is claimed that liquid hydrogen for a FCV could be filled in as little as 5 minutes (Ogden & Nitsch, 1993), assuming of course that widespread cryogenic filling stations would be installed. Similar refilling requirements would be required for carbon adsorption systems for which a five minute filling time is projected. Finally, metal hydrides storage is projected to have the longest filling time for hydrogen at 20–30 minutes, with the extra time being required for (exothermic) chemical reactions to take place as the hydrogen penetrates the crystalline structure of the medium.

Finally, regarding operational obstacles to electric vehicles, start-up times are inordinately long for some of the present technologies (Ogden & Nitsch, 1993). Presently available fuel cells, such as the phosphoric-acid and alkaline types, require anywhere from 2 to more than 5 hours to bring them to their operating temperatures in the range 80–200 °C (176–390 °F). PEM fuel cells, by comparison may reduce the warm-up time to about 5 minutes. The PEM cells, which are still under development, achieve this by allowing operation at lower temperatures (120 °C (248 °F) and below). The high-temperature batteries, such as the sodium–sulfur cell, also will require a warm-up time to be operational from cold initial conditions.

Ambient-temperature batteries, of course, deliver their energy immediately upon connection to the traction motors of an EV.

In closing our discussion of hydrogen FCVs, recent studies (Ogden & Nitsch, 1993) give an optimistic economic projection indicating that life-cycle costs for these vehicles can be made comparable to those of conventional ICE vehicles. The costs included in the comparisons were: fuel, fuel storage, fuel cell/motor, batteries, maintenance, and the vehicle itself. The FCV, with higher initial costs than the ICE, showed a slight advantage when capital costs were levelized at a low interest rate (6%) and only a marginal advantage when using a higher rate (12%). The life-cycle costs for both technologies, as measured per distance traveled, came out at approximately $0.035/mile ($0.022/km) (exclusive of taxes). The reasons cited for these surprising results were the much higher fuel efficiency of the FCV, the longer lifetime of the FCV, and the lower FCV maintenance costs. While these cost projections are those of hydrogen advocates, the factors appear to be realistic, even if quantitatively they might err somewhat on the side of optimism.

5.6 Thermal storage

Thermal energy storage has its principal application for sustainable technologies with solar heat-collecting systems. The collection of solar energy in the form of heat, be it for domestic hot water, space heating/cooling, thermal generation of electricity or whatever, is subject to diurnal cycles, seasonal shifts, and random weather variations. Buffering of the fluctuations of thermal outputs over short times is requisite for most solar-heat systems and is achieved relatively easily and inexpensively. Time-period shifting in daily operation, as mentioned earlier for electric storage, has also been used in prototype thermal operations. Longer-term heat storage, however, presents technical and cost challenges, as we shall see.

Thermal energy storage can, in principle, be accomplished simply by containing the heat in a material medium, whether liquid, gas, or solid. The most common storage medium for lower-temperature solar heat is, of course, water. Water is the storage medium for most SDHW systems and many (active) solar space-heating systems (Garg *et al.*, 1985; Winter, 1990; Hof, 1993). Water storage operates, of course, only at temperatures below the boiling point of water (unless pressurized), thus limiting the heat capacity and heat-transfer capabilities of the system. Solids, such as rock or metals, can also be used and will operate over a wider range of temperatures. Porous beds of pebbles, for example, are used for storage in hot-air solar systems for space heating. For storage of heat from working fluids at high temperatures, such as those used with solar central receivers (p. 36), molten salts or liquid metals (e.g. sodium) have been used.

Water and rock are examples of sensible-heat media in which the amount of heat energy stored is proportional to the rise in temperature, while the medium retains its physical state (e.g. liquid or solid) (Jensen & Sorenson, 1984). For any such medium, a specific heat can be defined that gives the amount of energy contained per unit of mass for each degree of temperature rise. Water has the highest specific heat of any of the liquids or solids usable for sensible heat storage and, historically, has served as the standard for definition of units of heat content in both the metric and the Imperial systems of units. For example, using either calories or British thermal units, as units of heat content, specific heat is defined, respectively, as either kilocalories/gram-degree Celsius or British thermal units/pound-degree Fahrenheit. The specific heat of water is, accordingly, equal to 1.0 in either unit system.

By comparison, varieties of rock have specific heats only about one fifth that of water and thus require a larger mass of the material to store an equivalent amount of heat. Rock, as used for hot-air storage, is in the form of pebbles with air voids between the pebbles; this reduces the mass density from that of solid rock (nearly three times the density of water). The net result of the lower specific heat of the individual pebbles and the density reductions of the air voids is that a pebble bed must be over twice the volume of water to achieve the same total heat capacity (per degree of temperature change). For example, to provide 1 day's heat storage for space heating of an average home would typically require a water tank of about $10m^3$ volume (2640 gallons or 9990 liters) (Swet, 1987), whereas a hot-air pebble bed would require over $20\,m^3$ (the volume of a cube approximately 2.7 m (6 ft) on a side for the pebbles compared with a bit over 2 m per side for the water). Since space is at a premium inside the average home, this larger volume would be a significant disadvantage for the pebble bed.

Latent-heat compounds are alternatives to sensible-heat media if higher heat-storage capacities are desired. These materials undergo a change in state in the process of absorbing or releasing heat (Garg *et al.*, 1985). The most commonly known change in state is water, when it is heated beyond its boiling point and changes from liquid to steam (a vapor). In making this change of state, an additional amount of heat, called the heat of vaporization, must be absorbed. Unlike sensible-heat changes, there is no accompanying change in temperature during this change of state from water to steam. Latent heat is also required to change a solid to a liquid. Common examples of these solid/liquid media are paraffin wax, Glauber's salt, and sodium metal, all with melting points below the boiling point of water. The heat exchange for the solid–liquid state change is called the heat of fusion.

These media are also called phase-change materials (PCMs). Because of the energy storage in heats of transition, PCMs offer possibilities of more compact storage, requiring a smaller mass of storage medium. Some PCMs also have advantages of very good thermal insulation and operation over wide ranges of temperature (Winter, 1990). Their use has been limited to trials in a few prototype solar projects, however. The characteristics of PCMs, such as their chemical stability under repeated thermal cycling, have not been fully investigated to date.

At present the PCM systems do not appear to offer cost reductions for storage in the lower temperature range (50–100 °C or 122–244 °F). PCM system costs, including structure, plumbing and the medium itself have

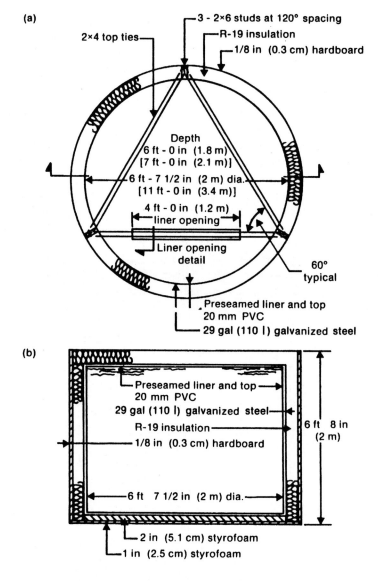

Figure 5.3
A low-cost hot-water storage tank: (a) plan; (b) section. (Adapted from Bourne, 1981.)

been projected to cost over twice those of hot-water systems. While the use of organics, such as paraffin or naphthalene, may have some operation and maintenance advantages in terms of low corrosion and chemical stability (Kilkis & Kakac, 1989), these gains would not seem to offset the higher investment costs. In addition, flammability and chemical instability at higher temperatures could be operational disadvantages. Finally, these materials themselves will always be more expensive than water and stone, against which they must compete, as material costs in the initial investment.

It should be recognized, however, that the material cost of the storage medium is not always the determining factor for the costs of thermal storage. Construction costs, including structural aspects, plumbing, and controls, must also be included (Winter, 1990) and often constitute the major cost of the storage system. As with solar-collector systems, low cost is essential for the market competitiveness of the solar energy delivered and great emphasis must be put on inexpensive fabrication. An example of a low-cost, hot-water storage tank is given in Fig. 5.3, showing a simple wood (hardboard) structure with inexpensive styrofoam insulation and a plastic membrane liner.

Estimates have been made of the unit costs of hot-water thermal storage. Swet (1990) has indicated that a steel water-storage tank should have a unit-volume construction cost of about $100/ft^3 ($3500/m^3) This would result in a unit-energy investment cost of about $900/MBTU ($853/GJ) storage capacity for a temperature change of 50 °F (28 °C) in sensible-heat storage or double that cost for a 25 °F (14 °C) swing. For diurnal storage service, such an initial cost would result in a levelized unit cost of delivered heat to be in the neighborhood of $1/MBTU, depending on the temperature swing, the number of annual storage cycles, the discount rate, and the amortization period (or storage tank lifetime). Such a cost of stored energy, as with the other forms cited earlier, must be added to the source costs. For the case of SDHW approximately $1 must be added to the levelized cost estimate of $17/MBTU ($16/GJ) for the collector system (p. 24), thereby making only an incremental increase over the collector costs (which are currently high compared with conventional fuel-heat costs (Al-Baharna, 1987; Winter, 1990)).

It should be realized at this point that this cost of delivered stored energy is based on the assumption of diurnal service. Typically, such service entails short-time storage, either 1–3 hours for buffering purposes or 4–5 hours for period displacement. The storage system is cycled daily or almost daily giving 250–350 cycles annually. This would not be the case for long-term storage, such as supplying the load over cloudy-weather sequences or seasonal storage, where the system is cycled fewer times annually. Since the (fixed) unit cost of energy delivered from storage must be inversely proportional to the number of annual storage cycles, there has to be a rise in the unit ($/MBTU) of storage as these cycles are lowered below daily use.

Research has been conducted into operations using large heat-storage capacities for carryover purposes such as for "seasonal storage" or for increasing the annual solar fraction (see p. 22) in diurnal operation (Swet, 1990; Veziroglu & Takahashi, 1990). In this type of operation, there is an attempt to carryover earlier collected solar heat for later use. The extreme case of carryover is seasonal storage, where summer heat is carried over to winter (or winter "cool" is carried over to summer). In any of these schemes, it is found that the marginal cost of energy delivered from storage increases as the ratio of storage capacity to collector area increases, when storage goes beyond the point of 2 to 3 days of collected heat. Since increasing the storage-to-collector ratio to the higher range gives diminishing returns in energy delivered, any attempt to approach a unity solar fraction will result in storage costs much higher than those for ordinary diurnal service with buffering storage only. Some of the research results suggest that unity-solar-fraction storage costs will be an order of magnitude higher than the costs under diurnal operation with back-up heat. The only application where such costs could be justified would be remote, stand-alone service where a back-up source is unfeasible.

High-temperature heat storage for large solar-thermal systems (such as the central receivers reviewed on p. 36) require other media, such as molten salts and special oils (Holl & De Meo, 1990). A schematic showing the integration of thermal storage into the overall system of the Solar One central receiver is given in Fig. 5.4. The storage medium in Solar One was a combination of oil and rock, operated at 304 °C (579 °F) with a storage capacity of 4 MW-hr (thermal). This capacity represents a storage time of only about 5 minutes (for full-rated thermal output power), which obviously serves only for buffering purposes. Later versions of storage for Solar One used molten salts as the medium and had a somewhat higher capacity (7 MW-hr). The storage costs for this early prototype central receiver were only about 8% of the total plant construction costs (De Meo & Steitz, 1990), because of the small role which storage played in the operation.

Most of the prototype models of the trough-collector SEGS projects (Lotkin & Kearney, 1991) relied on back-up burners instead of storage for smoothing or availability. On one of the later projects, however, high-temperature oil was used for short-term storage. The initial cost of this storage was estimated at about $25/kW-hr (de Laquil, et al., 1993). The added levelized cost of electric energy as delivered from storage in this case turns out to be less than 1¢/kW-hr, assuming daily cycling, a low interest rate, and a reasonably long plant life. Here again, it should be recognized that storage is playing a minor role in the operation of these solar–steam,

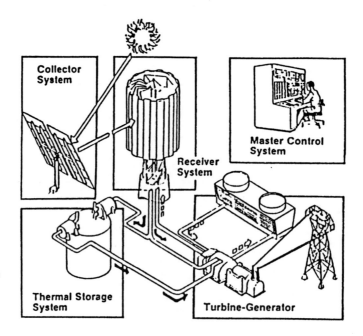

Figure 5.4
Thermal storage for the Solar One Central Receiver. (Adapted from Holl & De Meo, 1990.)

trough-collector plants, as it had for the central receiver plants mentioned above.

In future prototype central receiver plants, molten salts are planned for the working fluid as well as for the storage medium. The Solar Two project will use the molten salts as the storage medium operating at 566 °C (1051 °F), with sufficient storage capacity to supply the 10 MW$_e$ output for 4 hours after the sun goes down. The project is, therefore, claimed as "dispatchable" (DOE, 1993) to its load, the SCE grid. The costs of this storage will clearly form a significant fraction of the overall cost of the project.

For solar-thermal electric generation, the limited role of thermal storage restricts the penetration of the technology into the electric-system market just as the limits on electric storage do for solar–PV and wind. For other applications of solar-thermal receivers or trough systems, such as agricultural or industrial heat applications (Anon., 1992; DOE, 1992b), the lack of long-term heat storage may not be as restricting. In any application where the product of the heat operation may be stored, rather than storage of heat itself, this situation would obtain. Solar-pumped irrigation is an example where water pumped during sunny periods can be stored much more easily and inexpensively than storing the solar heat itself. (The same would apply, of course, to PV or wind generator irrigation pumping with stored water versus stored electricity.) Crop drying is another thermal application in agriculture, where the old farming adage "make hay while the sun shines" comes to mind.

IPH is the other area of opportunity for solar-thermal heat (Brown, 1983; Anon., 1992; DOE, 1992b) where storage of the product might be considered. The processing products are many, including paper and food products, which may be accumulated as the processing energy is available. Here, however, there are going to be limitations on the variety of products that can be stored, in consideration of the requirements of the manufacturing process itself. Such considerations would have to account not only for the material properties in processing but also for the impact of interruptions of process heat on the flows of the operation. A better example for purely thermal processes is solar-thermal desalinization, where the processed (fresh) water is stored in the same manner as ordinary reservoir water is.

Solar ponds (p. 44) provide their own storage, at least for buffering and period-shifting purposes. Solar ponds have been used for thermal–electric generation (p. 44) and for space heating (Watson, 1989). Since the heat-trapping density gradients are an integral part of solar-pond operation, storage exacts no added costs to an already low-cost installation compared with other means of solar electric generation. Solar ponds offer limited possibilities for large-scale adoption, however, because of their regional (sunbelt) operational restrictions and their scale limitations.

Perhaps the best examples of energy storage as integral elements of the overall design are found in passive solar architecture (p. 30). Various ingenious means of incorporating walls, floors, or ceilings that absorb sunlight in the daytime and reradiate the heat in the evening, with few or no moving parts, have been conceived and executed in passive solar

(Anderson, 1990) or bioclimatic (Watson, 1989) designs. Trombe walls (water filled) have long been successfull and the more recent "thermic diode" cell is a clever innovation (Fig. 5.5) for passive means of accomplishing the period-displacement function discussed here for several different active solar technologies. Still other designs, such as the thermo siphon means of moving solar-heated water to storage, accomplish the same functions as active systems, even though they are passive in principle. All of these means of heat storage in solar architecture are components in highly cost-effective overall designs, but the costs of storage are not easily separated out as they were in the active technologies.

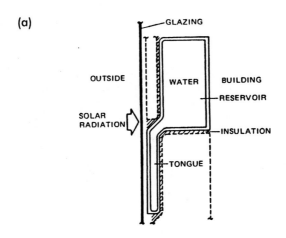

Figure 5.5
Thermic diode cell, for passive solar architecture. (a) Liquid convector diode. (b) Fluid behavior. (Adapted from Jones, 1984.)

5.7 Superconducting magnetic energy storage

Electrical energy can be stored directly in magnetic coils. Magnetic coils are simply coils of wire, often made in a solenoidal configuration, through which electric current flows. When the current flows through the turns of the coil, it sets up a magnetic field, just as found around a permanent magnet. One of the unexpected findings of 19th century exploration of electromagnetism was that energy is stored in the magnetic field surrounding such a coil when current is forced to flow through its windings. This stored energy is returned to the electric circuit by the collapsing magnetic field when the current is diminished or switched off. This is the physical principle of energy storage in a magnetic coil.

Energy storage in coils made of ordinary conductors is not practical because of resistance losses in the wires. Too much of the electric energy input would be converted to unusable heat with such conductors. The discovery of superconductivity in 1911 (Dahl, 1992) changed the thinking of engineers about the transmission and storage of electricity with the new possibility of using wires with zero resistance. Using superconducting wires they saw 100% transmission over power lines and retrieval of stored electric energy without any loss. Furthermore, for magnet coils, currents in superconductors are persistent, meaning that once established they continue to flow without a driving force (voltage) as long as the superconducting state is maintained.

But while superconductivity had been demonstrated in the laboratory, there were formidable practical difficulties to its use. The superconducting materials first identified, such as the metals tin and lead, turned to the zero conducting state only below a critical temperature near absolute zero. Such temperatures could only be attained by elaborate cryogenic techniques in a bath of liquid helium, and clearly such provisions were prohibitively expensive for ordinary operation of power systems. In addition, true zero resistance only occurs for d.c. currents in superconductors.

In the late 1980s, great excitement followed the discovery of a new class of material that went into the superconducting state at significantly higher temperatures, in the range of the boiling point of liquefied nitrogen (Dahl, 1992). These temperatures, around 77K (−196 °C or −353 °F), are much more easily and less expensively produced in operating systems. Therefore, all of the earlier expectations for zero-loss applications were aroused by the advent of this obscure group of ceramic oxides which almost inexplicably had these properties. The theoretical understanding of superconductivity, based in quantum mechanics, has from the start never reached a status of consensus in research circles (*Physics Today*, 1991; *Science*, 1993, 1994, 1996).

After the initial excitement amongst researchers over these "high-temperature" superconductors, came the realization that considerably more R&D would be necessary before there could be a technology feasible for applications such as power-system energy storage: superconducting magnetic energy storage (SMES). Not only would the critical temperature of the superconductor be a major factor for its applications, but also the strength of the magnetic field it is immersed in and the amount of current flowing in the conductor were found to be crucial for maintaining the superconducting state. When current densities (amperes per square centimeter, within the conductor cross-section) or magnetic field strengths go above certain critical values, superconducting properties are lost. Therefore, the research quest became to find materials that have sufficiently high critical values of current density and magnetic fields in addition to having a high critical temperature (Moore, 1992).

Research efforts in finding new materials and techniques for increasing the critical values of current density and magnetic field have made progress in recent years (*Science*, 1993, 1995). Laboratories in the USA, Japan, and France have come up with new versions of the exotic ceramic oxides (e.g. yttrium barium oxide) that can be operated in the superconducting state at temperatures in the range 47–77K (−407 to −353 °F), immersed in magnetic fields and carrying currents sufficiently high to be in the range required for the intermediate-size SMES applications. The fabrication of these ceramics into the wire lengths required and their durability under the stresses of operation, however, remain a daunting challenge for the development phase of superconductor R&D.

For SMES applications, long lengths of superconducting wires would be required in order to wind coils of the many turns needed to store the large amounts of electrical (magnetic) energy required for power applications. An appropriate amount of energy storage for the evolving sustainable technologies would be in the range 10–20 MW-hr, which would be adequate for storage operations at larger solar fields or wind farms. Such a range of energy-storage capacities has been termed an "intermediate size" of SMES plant in recent development efforts for electric utilities (Lubell & Lue, 1995) and would require coils of hundreds of turns, using thousands of meters of superconducting wires. At present, such lengths of the low-temperature, high-field, high-current superconductor wires have not even been approached in single strands and, therefore, (non-superconducting) splices would be required to obtain the required number of turns for such applications.

Engineering development projects have been undertaken for SMES systems, that could lay the ground work for power applications using the new superconductors if and when they are ready. These projects are based on designs using the existing (low-temperature) superconductors in liquid-helium baths (Lieurance *et al.*, 1995; Luongo *et al.*, 1995) and have power applications achievable in the short-term future. One such application is for power-system stabilization after momentary (less than a second) interruptions, where magnetic coils can supply energy more quickly than batteries or standby generators (De Winkle & Lamoree, 1993). In these uses, however, relatively small amounts of stored energy (e.g. 1 MW-hr) are needed and development projects for "micro-SMES" systems have been started. One such system has been operated in the field, having a small energy-storage capacity (about 0.2 KW-hr) but able to deliver electricity at a rate of nearly 0.3 MW for 2.5 seconds to provide the needed back-up for momentary outages.

Design projects have been undertaken for large-scale SMES systems (Luongo *et al.*, 1995), even with no definite prospects of when they might be adapted to the new superconductors. Such SMES systems – in the 5000 MW-hr, 1000 MW range – would have applications in utility load-leveling operations to replace conventional pumped-hydro storage plants. In load leveling, energy is stored during off-peak hours in sufficient amounts to meet the grid-system demands during peak hours. Since this demand is for the entire grid, load-leveling energy-storage capacities in the gigawatt-hour range are generally required. The coils for such storage capacities would be enormous, weighing hundreds of thousands of tons, and require earth support not only for their weight but also for the tremendous magnetic forces that would be exerted on the coils and their support structure. There are also safety concerns with such a massive operation, with exposure of personnel to extremely high (stray) magnetic fields and with the violent prospects of catastrophic loss of the superconductivity in the coils. The prospects for SMES systems of such capacities are long-term and highly uncertain.

The costs of superconducting energy storage, even in the intermediate-size range, can only be estimated since, at present, plants have only been built in the micro-SMES range (1–10 KW-hr) (De Winkle & Lamoree, 1993). Estimated unit capital costs for a 1 MW-hr demonstration plant (Lieurance *et al.*, 1995) are over two orders of magnitude higher than those cited earlier in this chapter for batteries. While capacity costs for SMES are said to scale as the square root of the storage capacity (Lubell & Lue, 1993), a 100 MW-hr intermediate-size plant would still be over an order of magnitude more in unit costs than batteries if this scaling holds true. This suggests than the prospects for SMES plants being cost-competitive with batteries, for use with the intermittent sources such as wind or solar, are not very promising within the foreseeable future.

Unforeseen developments could come from basic research in high-temperature superconducting materials and the underlying area of physics research in "condensed matter" (*Physics Today*, 1991; *Science*, 1993–96). Even if they do, however, extensive development efforts will be required to have an operable and economically competitive technology that could enhance the market penetration of the intermittent source technologies of interest in this book.

5.8 Summary

Energy storage offers limited possibilities for enhancing the adoption of sustainable energy-source technologies in the short term because of the present state of the energy-storage technologies themselves. We have found limitations in technological performance of batteries for mobile storage and non-competitive costs for stationary storage. While some prospects appear possible for innovations in the medium-term future through R&D, dramatic gains are not foreseeable at this time. Therefore, while the developments forecast in battery performance, because cost reductions, hydrogen-storage costs, or long-term heat-storage costs are conceivable, the outcomes are uncertain. Flywheels may turn out to be an unexpected alternative to the seemingly intractable problems of batteries, but this will become clear only after further R&D. Other possibilities, such as superconducting magnets usable for stationary storage of generated electricity, are nowhere near the technological development stage.

The consequences of these limitations, for the medium term at least, are several. The technological limitations on battery storage appear to confine the role of EVs, in the absence of major policy initiatives, to light-duty, short-range service for the near to medium-term future. Similar performance limitations persist for hydrogen FCVs, the other environmentally benign automotive prospect. Major cost subsidies are likely to be required in order to widen this limited adoption of electric vehicles or hydrogen FCVs required by environmental policies in the near term. In the meantime, interim (partial) solutions to urban air pollution have been proposed using hybrid EV/ICE vehicles to cover the long-distance requirements outside urban areas, where emissions can better disperse.

For stationary applications, short-time storage only seems feasible, whether for electricity or heat in the near-term future, meaning that high-availability solar or wind service does not appear to be likely within such a time span. Consequently, many stand-alone operations using solar and wind sources will continue to require a conventional back-up source and grid operation will be of limited penetration, with limited capacity factors. Storage for the solar sources (PV or heat) seems to be limited to buffering and period-displacement functions only in either stand-alone or grid applications for the foreseeable future.

Battery storage for wind generation, however, appears to have better prospects in the medium term because of more confident projections of combined generation/storage costs falling to the competitive range. This could lead to increased availability of energy derived from wind, higher capacity factors, and higher capacity credits for wind installations. All of these factors lead to enhanced investment prospects and improved operational characteristics, encouraging further (still partial) penetration into the utility-grid market.

References

Al-Baharna, W.S. (1987). Comparison of cost per GJ of insolation, solar water heating, electricity and other conventional fuels. In: *Alternate Energy Sources VIII*, Vol. 2, ed. Veziroglu T.N., pp. 881–886. New York: Hemisphere.

Anderson, B. (ed.) (1990). *Solar Building Architecture*. Cambridge, MA: MIT Press.

Anderson, M.D. (1993). Battery energy storage technologies. *Proceedings of the IEEE*, 8, 475–479.

Anon. (1992). Solar process heat technologies begin to make their mark. *Industrial Energy Technology*, Nov.

Barber, K.F. & Abarcar, R.B. (1992). A US Government/industry program to develop and commercialize electric vehicles. In: *Proceedings of the EVS – 11*, Sept., Paper 1.01, pp. 1–12.

BMI (Battelle Memorial Institute) (1995). *Clean Fleet – Final Report*. Columbus, OH: Battelle Memorial Institute.

Brown, K.C. (1983). Re-examinimg the prospects for solar industrial process heat. *Annual Revieew of Energy*, 8, 509–530.

Brown, S.F. (1995). Its the battery, stupid. *Popular Science*, Feb. (as cited in *International Journal of Hydrogen Energy*, 21, 317).

Bourne, R.C. (1981), *Membrane-lined Foundations For Liquid Thermal Storage*, DOE/ET/2011–1. Washington, DC: US DOE.

Burke, A.F., Hardin, E.J. & Dowgiallo, E.J. (1990). Application of ultracapacitors in electric vehicle propulsion systems. In: *Proceedings of the 34th International Power Sources Conference*, pp. 328–333. New York: IEEE.

Cassedy, E.S. & Grossman, P.Z. (1990). *Introduction to Energy – Resources, Technology and Society*. Cambridge, UK: Cambridge University Press. Appendix A: Thermodynamic definitions (pp. 269–271)and Appendix C: Advanced storage (pp. 322–326).

CEC (California Energy Commission) (1991). Cost and availability of low-emission motor vehicles and fuels. *AB234 Report* (Draft, Aug. 1991). P500-91-009. Sacramento, CA: California Energy Commission.

Chan, C.C. (1993). An overview of electric vehicle technology. *Proceedings of the IEEE*, 81, Sept, 1202–1213.

Chowdhury, B.M. (1988). Is central station photovoltaic power dispatchable? *IEEE Power Engineering Review*, Dec., 30.

Chowdhury, B.M. (1991). Optimizing the integration of photovoltaic systems with electric utilities. In; *Proceedings of the IEEE PES Summer Meeting*. Paper 91 SM329–3EC.

Conway, B.E. (1991). Transition from supercapacitors to battery behavior in electrochemical energy storage. *Journal of the Electrochemical Society*, 138, 1538–1548.

Culta, M. (1989). Energy storage systems in operation. In: *Energy Storage Systems*, eds. Kilkis, B. & Kakac, S., pp. 551–574. Dordrecht, the Netherlands: Kluwer Academic.

Dahl, P.F. (1992). *Superconductivity – its Historical Roots and Development from Mercury to the Ceramic Oxides*. New York: American Institute of Physics.

de Laquil III, P., Kearney, D., Geyer, M. & Diver, D. (1993). Solar–thermal electric technology. In: *Renewable Energy – Sources for Fuels and Electricity*, Ch. 5, eds. Johansson, T.B., Kelly, H., Reddy, A.K.N. & Williams, R.H., pp. 213–296. Washington, DC: Island Press.

De Meo, E.A. & Steitz, P. (1990). The US electric utility industry's activities in solar and wind energy. In: *Advances in Solar Energy*, Vol. 6, ed. Boer, K.W., pp. 1–218 Boulder, CO: American Solar Energy Society.

De Winkel, C. & Lamoree, D.J. (1993). Storing power for critical loads. *IEEE Spectrum*, June, 38–42.

Dickinson, E.M., Ryan, J.W. & Smulyan, M.H. (1977). *The Hydrogen Energy Economy – A Realistic Appraisal of Prospects and Impacts*. New York: Praeger.

DOE (US Department of Energy) (1991). *Photovoltaics Program Plan. FY 1991–*

FY 1995. DOE/CH10093–92. Golden, CO: National Renewable Energy Laboratory.

DOE (1992a). *Hydrogen Program Plan. FY 1993–FY 1997*. DOE/CH 10093–147. Golden, CO: Office of Conservation & Renewable Energy, National Renewable Energy Laboratory.

DOE (1992b). *Solar Industrial Program. Program Plan FY 1993–FY 1997*. (Draft, 9 December, 1992). Washington, DC: Office of Industrial Technologies, Office of Conservation & Renewable Energy.

DOE (1993). *Solar Thermal Electric, Five Year Program Plan, FY 1993–FY 1997*. Washington, DC: Solar Thermal and Biomass Power Division, Office of Solar Energy Conversion, (G.D. Burch, Director).

Dowgiallo, E.J. & Harding, J.E. (1995). Perspective on ultracapacitors for electric vehicles. *Proceedings of the 10th Annual Battery Conference on Applications and Advances*, pp. 153–157. New York: IEEE.

EPRI (Electric Power Research Institute) (1991). The push for advanced batteries. *EPRI Journal*, April/May, 16–19.

Ford, A. (1994). Electric vehicles and the electric utility company. *Energy Policy*, 22, 555–570.

Garg, H.P., Mullick, S.C. & Bhargava A.K. (1985). *Solar Thermal Energy Storage*. Dordrecht, the Netherlands: Reidel.

Genk, G. (1985). *Kinetic Energy Storage – Theory and Practice of Advanced Flywheel Systems*. London: Butterworth.

Gregory, D.P. (1973). The hydrogen economy. *Scientific American*, Jan. 13–21.

Grubb, M.J. & Meyer, N.E. (1993). Wind energy: resources, systems and regional strategies. In: *Renewable Energy – Sources for Fuels and Electricity*, Ch. 4, eds. Johansson, T.B., Kelly, H., Reddy, A.K.N. & Williams, R.H., pp. 157–212. Washington, DC: Island Press.

Hof, G. (1993). *Active Solar Systems*. Cambridge, MA: MIT Press.

Holl, R.J. & De Meo, E.A. (1990). The status of solar thermal electric technology. In: *Advances in Solar Energy*, Vol. 6, ed. Boer, K.W. Boulder, CO: American Solar Energy Society.

Humphreys, K.K. & Brown, D.R. (1990). Life-cycle cost comparisons of advanced storage batteries and fuel cells, stand-alone and electric vehicle applications. *Report PNL–7203, UC–212*. Richland, WA: Battelle Pacific Northwest Laboratory.

IAHE (International Association for Hydrogen Energy) (1990). Hydrogen energy progress. In: *Proceedings VIII of the 8th World Hydrogen Energy Conference*. Oxford, UK: Pergamon Press.

IEA (International Energy Agency) (1993). *Electric Vehicles: Technology, Performance and Potential*. Paris, France: Organization for Economic Cooperation and Development.

Ingersoll, J.G. (1991). Energy storage systems. In: *The Energy Sourcebook – A Guide to Technology, Resources and Policy*, eds. Howes, R. & Fainberg, A. New York: American Institute of Physics.

Jefferson, C.M. (1993). Power management in electric vehicles. In: *IEE Colloqium for Electric Racing Vehicles*. London, UK: Institution of Electrical Engineers.

Jensen, J. & Sorenson, B. (1984). *Fundamentals of Energy Storage*. New York: Wiley.

Jones, G.F. (1984). Liquid convective diodes. *Proceedings of the National Passive Solar Conference: Passive and Hybrid Solar Energy Update*, Sante Fe, NM.

Kelly, H. & Weinberg, C.J. (1993). Utility strategies for using renewables. In: *Renewable Energy – Sources for Fuels and Electricity*, Ch. 23, eds. Johansson, T.B., Kelly, H., Reddy, A.K.N. & Williams, R.H., pp. 1011–1069. Washington, DC: Island Press.

Kilkis, B. & Kakac, S. (eds.) (1989). *Energy Storage Systems.* Dordrecht, the Netherlands: Kluwer Academic. (a) Culta, M., Energy storage systems in operation, pp. 551–574; (b) Culta, M., Superconducting magnetic energy storage , pp. 575–597; (c) Kilkis, B. & Kakac, S., Importance of energy storage, pp. 1–10.

Kordesch, K. & Oliveira, J.C.T. (1988). Fuel cells: the present state of the technology and future applications with special consideration of the alkaline hydrogen/oxygen (air) systems. *International Journal of Hydrogen Energy*, 13, 411–427.

Lieurance, D., Kimball, F., Rix, C. & Luongo, C. (1995). Design and cost studies for small superconducting magnetic energy storage (SMES) systems. *IEEE Transactions on Applied Superconductivity*, 5, 350–353.

Lotken, M. & Kearney, D. (1991). Solar thermal electric performance and prospects – the view from LUZ. *Solar Today*, May/June, 10–13.

Lubell, M.S. & Lue, J.W. (1995). Structure and cost scaling for intermediate size superconducting magnetic energy storage (SMES) systems. *IEEE Transactions on Applied Superconductivity*, 5, 345–349.

Luongo, C. & the Bechtel SMES Team (1995). Review of the Bechtel team's SMES design and future plans for a technology demonstration unit, *IEEE Transactions on Applied Superconductivity*, 5, 422–427.

MacKenzie, J.J. (1994). *The Keys to the Car – Electric and Hydrogen Vehicles for the 21st Century.* Washington, DC: World Resources Institute.

McHarnon, F.R. & Cavins, E.J. (1989). Energy storage. *Annual Review of Energy*, 14, 241–271.

Miller, M. (1996). Fuel cell analyst cites PEM cell as most promising. *International Journal of Hydrogen Energy*, 21, 428–429.

Moore, S.T. (1992). Superconductors are still hot. *EPRI Journal*, Sept., 5–15.

Moore, T. (1993). Charging up for electric vehicles. *EPRI Journal,* June, 6–7.

NESEA (Northeast Sustainable Energy Association) (1997). *Tour de Sol News*, 97, July (US Electric Vehicle Championship, Greenfield, MA).

NESEA (1998). *Northeast Sun*, 16, Summer (whole issue).

NRC (National Research Council) (1976). *Criteria for Energy Storage R&D.* Washington, DC: National Academy of Sciences, Committee on Advanced. Energy Storage Systems.

Ogden, J.M. & Nitsch, J. (1993). Solar hydrogen, In: *Renewable Energy – Sources for Fuels and Electricity*, Ch. 22, eds. Johansson, T.B., Kelly, H., Reddy, A.K.N. & Williams, R.H., pp. 925–1009. Washington, DC: Island Press.

Physics Today (1991). Special issue: high temperature superconductivity. *Physics Today,* June, 22–82.

Post, R.F., Fowler, T.K. & Post, S.F. (1993). A high-efficiency electromechanical battery. *Proceedings of the IEEE*, 81, 462–474.

Science (1993). (a) New superconductors: a slow dawn, (Research News), 259, 306–308; (b) Holding the line in hightemperature superconductors, 261, 1521–1522. (c) High Tc superconductors get squeezed (Materials Science), 262, 31; (d) A big step for superconductors? (Research News), 262, 1816–1817;

Science (1994). (a) Superconductivity researchers tease out facts from artifacts, (Meeting Briefs), 265, 2014–2015; (b) Cuprate superconductors: a broken symmetry in search of a mechanism [Perspectives (S. Chakravarty)], 266, 386–387.

Science (1995). New superconductor stands up to magnetic fields (Materials Science), 268, 644.

Science (1996). Closing in on superconductors, Research News, 271, 288–289.

Stoll, H.G. (1989). *Least-cost Electric Utility Planning.* New York: Wiley.

Swet, C.J. (1987). Storage of solar thermal energy – status and prospects. In: *Progress in Solar Engineering*, ed. Goswami, D.Y. Washington, DC: Hemisphere.

Swet, C.J. (1990). Issues and opportunities. In: *Solar Collectors, Energy Storage and Materials*, ed. Winter de F., Ch. 19, pp. 796–827. Cambridge, MA: MIT Press.

Veziroglu, T.N. & Takahashi, P.K. (eds.) (1990). *Hydrogen Energy Progress VIII*, Vol. 3. New York: Pergamon Press.

Wald, M.L. (1994). California regulators to meet on electric cars. *New York Times*, May 12, D2.

Watson, D. (1989). Bioclimatic design research, In: *Advances in Solar Energy*, Vol. 5, ed. Boer, K.W. Boulder, CO: American Solar Energy Society.

Winter de F. (ed.) (1990). Solar collectors, energy storage and materials. Cambridge, MA: MIT Press. (a) Swet, C.J., Overview of energy storage for solar systems, Ch. 14, pp. 611–624; (b) Swet, C.J., Storage concepts and design, Ch. 15., pp. 675–691; (c) Swet, C.J., Issues and opportunities, Ch. 19, pp. 796–827.

Further reading

Abhat, A. (1980). Short term thermal energy storage. *Review of Physical Applications*, 15, 477–450.

Allen, T. (1993). *Assessment of battery technologies for electric vehicles.* London: Institution of Electrical Engineers.

Brown, D.R. & Russel, J.A. (1986). A review of storage battery cost estimates. *Report PWL–5741.* Richland, WA: Battelle Pacific Northwest Laboratory.

Hsu, C.S. & Lee, W.J. (1993). Superconducting magnetic energy storage for power system applications. *IEEE Transactions on Industry Applications*, 29, 990–996.

Hull, J.R. (1997). Flywheels on a roll. *Spectrum*, July, 20–25.

IEEE (1994). Section on spinning wheels, in article on transportation. *IEEE Spectrum* (Technology issue), Jan., 62.

Imarisio, C. & Strub, A.S. (eds.) (1983). on *The Proceedings of the 3rd International Seminar Hydrogen as an Energy Carrier., Lyon, France.* Dordrecht, Holland: Reidel.

Lasseter, R.H. & Jalali, S.G. (1991). Power conditioning for superconducting magnetic energy storage. In: *The Proceedings of the IEEE Power Engineering Society,* Winter Meeting, New York. Paper 91 WM 133–9.

Lloyd, R.J., Schoenung, S.S. & Nakamura, T. (1987). Design advance in superconducting magnetic energy storage. *IEEE Transactions on Magnetics*, 23, 1323–1330.

Lodhi, M.A.K. (1989). Collection and storage of solar heat. *International Journal of Hydrogen Energy*, 14, 379–411.

Moore, T. (1994). Producing the near-term EV battery. *EPRI Journal,* April/May, 6–13.

NESA (Northeast Sustainable Energy Association) (1998). *Northeast Sun,* 16, Summer.

van Vort, W.D. & George, R.S. (1997). Impact of the California Clean Air Act. *International Journal of Hydrogen Energy,* 22, 31–38.

Veziroglu, T.N. (ed.) (1987). *Alternate Energy Sources VIII*, Vol. 2. New York: Hemisphere. (a) Healy, H.M., Cost and performances of large volume solar water heating systems for commercial and industrial facilities in Florida, pp. 869–880; (b) Al-Baharna, W.S., Comparison of cost per GJ of insolation, solar water heating, electricity and other conventional fuels, pp. 881–886; (c) Sauer, W.J. & Wang, Y.C., Proposed handbook of energy costs of construction, pp. 887–891.

6 Geothermal energy

Geothermal energy is an alternative source of heat and electricity generation that is customarily listed along with the renewable energy sources. It is a relatively clean energy source, emits small amounts of carbon dioxide, and might appear to be inexhaustible. We would, therefore, be tempted to include it in our list of sustainable technologies, especially if it had prospects for being a major contributor to alternative energy production in the future. Unfortunately, the potential for geothermal energy is limited (regionally) and it *is* exhaustible (at any given site). Furthermore, technology advances, which are a key factor in our assessments of other renewable sources, do not appear to offer high hopes for changing these limited prospects for geothermal contributions; therefore, our review here will not be extensive.

Geothermal power is derived by tapping steam or hot water that has been heated underground by the close proximity of aquifers to intrusions of the earth's hot magma near the surface (Armstead, 1973; Di Pippo, 1979; Howes, 1991). Favorable conditions are found only in certain regions around the globe where tectonic motion in recent geologic times (25 million years) has allowed magma intrusions into the earth's crust. Such regions are, naturally, also those where there is volcanic activity, or was in recent geologic times. Within the USA (lower 48 states), exploitable geothermal sites are most likely in the far west, with the prospects for steam only around the fault line paralleling the Pacific Coast.

Exploitable resources also exist in Alaska and Hawaii. Potential geothermal resources exist along the Gulf Coast of Texas and Louisiana in "geopressurized dome" formations, but the required drilling technology at great depths and high pressures has never been developed. Elsewhere in the world, resources are found along the Pacific Coast of South America, along an east–west line from the Mediterranean across the Middle East into Asia, in the Western Pacific Islands, and in Iceland.

While thermal baths were used by the Romans and subsequently in Europe, geothermal power technologies did not begin until the 20th century (Di Pippo, 1979; Palmerini, 1993). The first production of electricity from geothermal steam was in Italy in the year 1904, but significant development elsewhere did not start until the middle of the 20th century. In the 30 years from 1950, over 1.7 GW_e aggregate generating capacity was built worldwide and this figure was more than doubled in the next decade. Some projections suggest close to 9 GW_e total worldwide capacity by the year 2000 (Hutter, 1990), with the larger installations occuring in areas outside the USA: the Philippines, Mexico, Italy, New Zealand, and Japan (Table 6.1). Even with these impressive growth figures, geothermal elec-

Table 6.1. *Worldwide installed and planned geothermal power capacity (MW)*

	1980	1990	1995	Future
China	0	20.8	50	
Costa Rica	0	0	110	
Dominica	0	0	10	
El Salvador	95	95	125	355
France	4.2	4.2	4.2	
Greece	0	0	?	
Guatemala	0	2	15	30
Iceland	41	44.6	110	
Indonesia	32.25	142.25	379	
Italy	459	545	885	
Japan	214.6	214.6	456.5	
Kenya	45	45	105	
Mexico	425	700	950	
New Zealand	167.2	283.2	300	
Nicaragua	35	35	?	
Philippines	891	891	2164	
Portugal (Azores)	3	3	?	
St Lucia	0	0	10	
Thailand	0	0.3	3	
Turkey	20.6	20.6	40	
USA	1444	2770	3170	
USSR (former)	11	11	81	
Total	3887.85	5827.55	8967.7	
Total excluding USA	2443.85	3057.55	5797.7	
Overall increase (%)		49.89	53.88	
Increase excluding USA (%)		25.11	89.62	

Source: Geothermal Resources Council (1990). The figures are based on responses to a 1990 questionnaire and on data published in 1985.

tricity generation in the mid-1990s is only about 0.2% of the world installed electricity-generating capacity and it will still form a small fraction of the world's electric capacity even if the growth projections to the end of the 20th century are fulfilled.

The USA is the leading producer of geothermal electricity, with close to 2.8 GW$_e$ installed capacity in 1990 (Hutter, 1990) and the largest single complex of plants ("The Geysers" north of San Francisco). California has over 90% of the developed geothermal capacity in the USA (Williams & Bateman, 1995). Only a few hundred megawatts of installation are projected by the industry for the USA to the end of the 20th century, although such low projections are the subject of some dispute (US GAO, 1994). The Geysers complex was started in 1960 by PG&E Co. and was designed for a 2.0 GW$_e$ output capacity to supply 6% of California's electric power. The output of these plants has dropped in recent years, as discussed in *Science* (Anon., 1991), because of a depletion of the aquifer water feeding the wells.

Steam pressures have fallen and available thermal power is expected to be only half the designed capacity within a few years. This illustrates the exhaustible nature of the resource, which is unable to renew itself at the rate at which it was being exploited.

The Geysers and most other geothermal plants are said to be of the hydrothermal type, meaning that natural aquifers supply the water to be heated by the magnum below (Armstead, 1973). It is also possible to supply this water artificially from the ground level or to recharge natural hydrothermal wells through separately drilled well pipes. Recharging, in fact, is being tried at The Geysers but so far is insufficient to reverse the declining output of those wells. Some attempts have been made to inject water into hot dry rock formations in an effort to make steam or hot water, but such projects are still in the experimental stages (Palmerini, 1993).

Another way to use geothermal energy is by direct use of the heat (steam or hot water), either for IPH or for space heating (Di Pippo, 1979; Golof & Brus, 1993). Iceland is a prime example of this use, with over 85% of the residences in the capital city of Reykjavik being heated by geothermal water. Iceland, incidentally, is situated on the Mid-Atlantic Ridge crustal rift, which intersects no other inhabited land areas where its heat could be utilized. Direct use of geothermal heat is made in over 25 other countries, with the largest users being China and Japan. The total thermal power capacity worldwide – for direct use – is close to 12 GW_{th} (about the rate at which heat is supplied to four major power plants). The aggregate of all direct uses at present constitutes only about 0.04% of the world total primary energy usage, however, and is projected to be only 0.07% by the beginning of the 21st century (Palmerini, 1993).

Besides resource limitations, geothermal power has operational and environmental problems that must be dealt with. Geothermal effluents carry corrosive salts and pollutant gasses. The salt build-ups require steady maintenance to keep them from clogging boiler tubes. In addition, the combination of salts and low-temperature steam requires steam turbine blades to be made of special alloys that do not require frequent replacement because of corrosion. The gases that are emitted after release from geological pressures underground include hydrogen sulfide, CO_2, and radon. The hydrogen sulfide must be scrubbed out in order to eliminate this explosive poisonous gas with its noxious "rotten eggs" smell. The disposal of toxic wastes from this scrubbing and the processing of geothermal fluids must also be done with care. If the geothermal fluids (mostly water) are not reinjected, then pollution of other usable ground water could occur.

In those sites where geothermal wells can be drilled at depths less than 4000 m (13 000 ft), economic production of heat or electricity can be achieved. The less the depth, the lower the investment costs and, therefore, the lower the cost of the extracted energy. Geothermal electricity costs reportedly are in the region of \$0.03–0.010/kW-hr and geothermal heat in the range \$1.60–\$2.60/MBTU (\$1.50–2.50/GJ) (Palmerini, 1993), the lower ends of which are quite competitive with fossil fuels. Consequently, this resource is a boon in the localities where it can be found. Worldwide producer revenues for geothermal electricity were over $2 B and direct-use heat had revenues of over $0.25 B, which are significant to those few localities that have the resource.

The prospects for more widespread availability of geothermal energy are not great and technology innovations do not seem to hold reasonable probabilities of changing this situation significantly in the foreseeable future. Additional prospects for the resource exist outside the known hydrothermal regions only in hot dry rock (HDR) formations or geo-pressurized zones (Palmerini, 1993), both of which are tapped only at great depths (many miles deep) at great expense, if at all. Geopressurized zones are suggested as huge potential resources for geothermal energy – as they have been for natural gas – particularly along the Gulf Coast of the USA, but daunting technical problems with high pressures and high salinity would have to be overcome even before costs could be addressed. Even more uncertain are proposals to tap into the magma chamber itself, where the least depth that this may be done is over 6 miles (10 km) (Golof & Brus, 1993). Further development efforts continue on all these areas, but prospects for exploitation of regions such as HDR formations do not appear likely for decades to come.

References

Anon. (1991). Geothermal tragedy of the commons (Research News). *Science*, 253, 134–135.

Armstead, C.H. (ed.) (1973). *Geothermal Energy*. Paris: UNESCO Press.

Di Pippo, R. (1979). *Geothermal Energy as a Source of Energy*. DOE/RA/28320–1. Washington, DC: US Department of Energy.

Golof, R. & Brus, E. (1993). Geothermal energy. *The Almanac of Renewable Energy*, Ch. 4. Henry Holt for World Information Systems.

Howes, R.H. (1991). Geothermal energy. In: *The Energy Sourcebook*, pp. 239–255. New York: American Institute of Physics.

Hutter, G.W. (1990). Geothermal electric power – a 1990 world status update. *Geothermal Resources Council Bulletin, 19,* No. 7. Davis, CA: Geothermal Resources Council.

Palmerini, C.G. (1993). Geothermal energy. In: *Renewable Energy – Sources for Fuels and Electricity*, eds. Johansson, T.B., Kelly, H., Reddy, A.K.N. & Williams, R.H., pp. 549–591.Washington, DC: Island Press.

US GAO (US General Accounting Office) (1994). *Geothermal Energy – Outlook Limited for Some Uses but Promising for Geothermal Heat Pumps*. GAO/ RCED-94-84. Washington, DC: US General Accounting Office.

Williams, S. & Bateman B.G. (1995). *Power Plays – Profiles of America's Independent Renewable Electricity Developers*. Washington, DC: Investor Responsibility Research Center.

Further reading

Brown, L.R., Kane, H. & Ayes, E. (1993). Geothermal power gains. In:

Vital Signs 1993 – The Trends that are Shaping Our Future. Washington, DC: Norton for World Watch Institute.

DOE (US Department of Energy) (1992). *Geothermal Energy Program Review.* DOE/CH10093–182. Golden, CO: National Renewable Energy Laboratory.

7 Ocean energy

7.0 Introduction

The term ocean energy conjures up images of immense power, with winds, waves, and vast expanses of open water. The world's oceans do, in fact, hold huge amounts of energy. The idea of converting some of this energy for use has intrigued people for centuries, but extracting significant amounts of energy has not yet been accomplished. There are three forms of ocean energy that have been considered for tapping in recent years: wave energy, tidal power, and ocean thermal energy, each of which will be reviewed briefly in the following.

7.1 Wave energy

Ocean waves carry energy that may be converted into useful mechanical motion or hydraulic pressure, either one of which can be used directly or be used to drive an electric generator. There are several physical principles upon which such conversion can take place (Sanders, 1991): through the "heaving" (up and down) motion of the waves, by means of hydraulic piston pressure, through "surging" motion (of breaking waves), and from underwater pressure variations. There are over 1000 patents in the industrial world for ocean-wave devices, starting from almost 200 years ago. Despite this long history, however, wave power is not in wide use today.

Perhaps the most prevalent type of use of waves for energy is with navigational buoys, where the hydraulic piston principle is used. In this device, the up/down motion of the buoy in the waves causes variations in a column of water located in a piston-like cylinder within the buoy. The oscillations of the water column, in turn, cause air-pressure variations in the piston that are used to drive a pneumatic turbine. The turbine is the prime mover for an electric generator that supplies the lights and other navigational aids of the buoy. There are thousands of such air-turbine-powered buoys in operation worldwide today.

The air-turbine buoy application is similar to the "remote-site" applications for photovoltaics: an effective application but not an indication that the technology is, at present a prospect for mass production of power. In fact, wave technology development for larger-scale use is nowhere as far advanced as the PV or wind technologies. There are small R&D efforts for wave conversion in several countries, including the Scandinavian countries, Great Britain, Portugal, Japan, and the USA (Cavanagh, 1993).

Most of these efforts have been government sponsored but some have been privately undertaken such as that at the EPRI in California. The trend of the development efforts has been to build prototype models based on the different principles, with generating capacities ranging up to 100 KW$_e$ and plans to go to several hundred kilowatts. It has been estimated that an aggregate of 6 GW$_e$ wave-generation capacity is technically feasible in Great Britain (ETSU, 1995). The pace and backing of these projects, however, would not suggest significant commercialization within the foreseeable future.

If and when commercialization is considered, how widespread the application could become will depend on wave-energy resources. Like solar and wind technologies, wave-energy utilization will only be economic at sites where the resource is abundant. It has been shown that a meaningful measure of wave-energy resources at particular sites, either on- or offshore, is the energy carried along by the wave in the direction of propagation per unit of length along the wave front (which is transverse to the direction of propagation). If the rate at which this energy is incident at a point (offshore or on the shoreline) is given in kilowatts, then the power per unit of transverse length may be given in kilowatts per meter. In the case where the waves are incident at right angles to the shoreline, then the transverse direction is along the shoreline and the wave power density is in kilowatts per meter along that shoreline. This power density will depend on sea conditions, which will vary in magnitude and duration with the seasons and the weather. Wave-resource data for any given coastline are, therefore, given in terms of annual averages of kilowatts per meter.

Economic assessments have been made of wave-power generators using estimated annual energy outputs of typical wave-converter devices for assumed annual average resource power densities and the estimated unit capital costs of the devices (Sanders, 1991; Cavanagh, 1993). The levelized cost of the energy generated was calculated from these assumptions using standard methods and moderate discount rates. The cost of energy was at a competitive $0.10/kW-hr or less only if the average annual power density of waves was 50 kW/m or greater. If the annual power density was 20 kW/m or less, the cost of energy would be $0.20/kW-hr or greater.

Sites are known along some North Atlantic coasts (e.g. England and Scotland) with average wave power densities over 50 kW/m. However, there are sites in the same general region (e.g. Norway) where the power density is 25 kW/m or less. In the southern hemisphere, promising coastal sites (e.g. Australia) are known, having averages of 50 kW/m. In general, the most promising sites will be found in the area between latitudes 40° and 60° north or south because the winds at sea (which drive the waves) are the strongest and most consistent there.

Wave-generation technology is still in an early stage of development, despite its long history. The R&D efforts throughout the world are relatively minor compared with those for other prospective energy technologies. Much must be learned about the design of coastal installations to withstand large storms and the cost impacts of adequately robust designs, which could keep the technology priced out of the market. Valuable information in this regard may result from the operation of the massive offshore oil and gas drilling platforms.

7.2   Tidal power

Electric power may be generated from the rise and fall of the tides. Tidal-power plants operate on the same hydraulic principles as ordinary hydro-electric plants but depend on the differences in heights of water between high tide and low tide, and the flows of the water of the ebb and flood tidal currents. In order to operate efficiently, however, these plants must be located in basins where tidal heights are much larger than in the open sea (Sanders, 1991). Such sites are found in bays and estuaries where the size and shape of the basin make a natural resonance with the tidal fluctuations, thus amplifying the motion of the water much as the sloshing of water in a bathtub.

The natural tidal basin is made into a generating site by the construction of a dam-like structure called a "barrage" at the basin's mouth, with sluice gates at an opening to provide a controlled flow of water. The plants are operated in a sequence synchronized to the tidal period, which is about 12 hours and 25 minutes between high tides. Starting at low tide, one set of sluice gates is opened to allow the flow of water in on the flood tide. At a point near high tide, when the basin has filled to its maximum height, these gates are closed. Then, after the tide on the ocean side has fallen to its lowest level (about 6 hours later), other sluice gates are opened leading to hydraulic turbines and generation commences with the maximum "head" (the water height of high tide above low tide). Generation can continue for 4–5 hours until the head is lost because of the drop of the water height in the tidal basin.

Since generation takes place only for 8–10 hours daily, capacity factors under such operation can hardly exceed 35%. This has been found to be the limit even with "double effect" operation, where some generation is done on the flood tide. The maximum height achievable at any given site varies with the phases of the moon, with the greatest during the spring tides of the full and new moons and the smallest during the neap tides of the half moons. The annual CF will, therefore, be averaged over these twice-monthly variations but will still represent generation for only about one third of the hours out of a year. As a consequence, the fixed (capital) charges of a tidal plant are distributed over far fewer annual hours of energy production than a conventional hydropower plant and, therefore, would have higher unit-energy costs (dollars per kilowatt-hour) than a hydroplant of the same capacity and cost.

Tidal plants have estimated capital costs that seem to run in a range that is higher ($1800–2300/KW$_e$) than those found for conventional hydro or fossil-fired power plants (Cavanagh, 1993). A major determinant in these costs is the size of the "civil works" (e.g. the barrages) and the long construction times required in a marine environment. Like hydroelectric projects, the cost of the energy produced by these projects is strongly dependent on discount rates. Accordingly, energy costs for two of the proposed tidal plants have been estimated at $0.094–0.110/kW-hr at an 8% discount rate and $0.116–0.130/kW-hr at 10%. Going by one assessment, tidal resources in Europe can be developed at a competitive $0.09/kW-hr only when discount rates are 10% or less.

Possible environmental impacts are mixed for this technology. On the one hand, tidal power *is* renewable and non-polluting. On the other hand, the massive construction of the barrage and associated civil structures is bound to be disruptive of boat traffic, fisheries, and, possibly, bird habitats. There is speculation that, once constructed, positive environmental impacts might accrue from tidal projects, such as redistribution of estuarial sediments to the benefit of fish and bird feeding grounds. But there are also likely to be unforeseen impacts as a consequence of changed tidal ranges and currents, as these may affect dispersal of effluents and turbidity and may shift sediment beds. Whatever the case may be, there has not, to date, been a detailed environmental impact study done for a tidal plant, existing or proposed.

There are only a few operating tidal plants worldwide today, the largest of which has a peak generating capacity of 240 GW_e. This plant is located on the LaRacine River in France, having a maximum tidal range of 8 m (26 ft) and an annual capacity factor of about 26% (Cavanagh, 1993). The plant operates with relatively low maintenance and has been available for operation 97% of the time. There are other tidal plants in operation but they are much smaller in capacity, with the next largest at 18 MW_e, located in Annapolis, Canada.

Potential tidal generating sites have been identified worldwide, including the British Isles, Western Europe, Russia, India, Australia, and the Americas. Some of the better known sites are: the Bay of Fundy (both Canadian and US sides), the Severn Estuary (UK), and San José, Argentina. Seven major proposed projects worldwide would have installed capacities totaling nearly 25 GW_e and would generate over 35 TW-hr electric energy annually, which would be a little over 1% of the world's generating capacity and about 0.3% of the world's electric energy. Thus, we can see that the impact of tidal power worldwide would not be major, even if all of these projects were built. In some regions such as Europe (Cavanagh, 1993) there could be a more noticeable impact, however, where tidal power could represent nearly 15% of installed capacity and generate about 4% of the electric energy.

7.3 Ocean thermal energy conversion

The world's oceans store enormous amounts of energy in the form of heat, since about a quarter of the solar radiation reaching the earth is absorbed into them. If this heat could be used in some way, there would be another superabundant energy resource just as solar radiation is itself. There are several difficulties with utilization of ocean heat, however. First, it has to be recognized that the resource is dispersed over vast areas and operation of any energy conversion technology in all seasons and weather presents daunting challenges. Second, any attempt to convert ocean heat into a useful form must face the limitations imposed by the basic thermodynamics of the small temperature differences to be found in sea water anywhere, as reviewed below.

Power plants, using the same thermodynamic principles as conventional fuel-fired power plants, can operate on the small differences of water temperature between the surface and the bottom of the ocean (Sanders, 1991). Whereas the temperature differences found in conventional, fossil-fired, steam plants are several thousand degrees – taken between the hot gases of the boiler and the cooling water of the condenser (Cassedy & Grossman, 1990, 1998) – the temperature differences for ocean thermal plants are far less. Ocean thermal energy conversion (OTEC) plants must function on differences of only about 20 °C (36 °F) and magnitudes even of this size are found only in tropical waters.

OTEC plants must use volatile compounds, such as ammonia, as working fluids instead of the steam/water working fluids of conventional power plants. These compounds pass from a liquid state into vapor when heated only 10 °C (18 °F) by the warm ocean water in a heat exchanger. The heat energy gained from the warm ocean water is then expended by the vapor as it turns the blades of a turbine, which itself is coupled to an electric generator. After coming out of the turbine, the vapor returns to the liquid state when it passes through a condenser cooled by water taken from the ocean bottom. The liquid working fluid is then put under pressure to enter the warm-water heat exchanger again in a classic thermodynamic closed cycle.

Even though the operation of a thermodynamic cycle on such small temperature differences is novel, thermodynamic principles tell us that the efficiency of energy conversion will be very low (Cassedy & Grossman, 1990, 1998). This follows from the principle that the maximum achievable conversion to usable energy is proportional to the temperature difference in the cycle. Given the small differences in temperatures between the warm surface waters used to heat the OTEC cycle, and the cool bottom waters used for condensing, actual conversion efficiencies of prototype plants have been in the region of 5%. This should be compared with 35% and higher for conventional power plants.

R&D programs leading to small demonstration facilities have been carried on in several countries in recent years, including France, Great Britain, India, Japan, the Netherlands, Sweden, and the USA. Most of the programs are government supported, but a few are privately sponsored, such as by Westinghouse. The DOE supported an OTEC program with annual funding levels reaching several million dollars by the late 1980s, but this was cut in steps in the early 1990s and then terminated in 1995 (DOE, 1992, 1995). Before termination of the US effort, a small working prototype was demonstrated in Hawaii with an electrical output of 15 kW$_e$. Construction of a 40 kW$_e$, open-cycle, onshore plant was started in 1993 with support from DOE and the State of Hawaii and was scheduled for demonstration in 1995 (Greer et al., 1995). Designs for larger prototypes in the output range 100–400 kW$_e$ have been produced but never built. Plants of this size, if successful, could make limited contributions to electric grids or serve isolated communities in some tropical, coastal regions.

The construction of even a modest size (100 MW$_e$) OTEC plant, however, would present huge engineering challenges (Cavanagh, 1993). If it is a floating plant, it would be a vessel with a displacement of 250000 ton (230000 mt) and would likely require mooring in 1000 m (3000 ft) of water. Its cold water pipe would have to be 20 m (65 ft) in diameter, extending to the ocean bottom and carrying over 100000 gallons per second of water. Unless an appropriate ocean site can be found, with deep water near a shoreline and with electrical load demand nearby, long lengths of undersea cable would be required to reach land. Finally, as with wave-energy generators, offshore OTEC plants must be able to withstand major storms and their costs would have to reflect robust designs. (Again, experience with offshore drilling platforms should be useful.)

If technical requirements for full-scale OTEC plants are achieved, cost will be the critical issue for commercial viability. Since experience to date has been limited to small prototypes, only estimates are available for full-scale plants on capital costs, operation, and maintenance. Estimated capital costs appear high, however, in the range $6000–10000/kW$_e$ (Sanders, 1991; Cavanagh, 1993). Therefore, even with low operation and maintenance costs (less than $0.01/kW-hr), the levelized cost of energy would reach values exceeding $0.25/KW-hr, using moderate discount rates. Such costs of electricity are not competitive, of course, in most regions of the world today.

With little development effort in progress, the prospects for OTEC technology moving closer to economic feasibility seem dim for the foreseeable future. Even with commercialization, OTEC plants would be restricted to particular sites in tropical regions. The prospects for major impacts on world electric generation seem remote at present.

7.4 Summary

Ocean energy sources do not appear to have the potential to make a major impact on energy markets worldwide. They may, like geothermal sources, come to have importance in local areas if there are further technological development and cost reductions. It is with these prospects in mind that these technologies have been given less treatment here, since our emphasis is on those sustainable sources that can make major displacements of fossil fuels.

References

Cassedy, E.S. & Grossman, P.Z. (1990, 1998). *Introduction to Energy – Resources, Technology and Society,* 1st & 2nd edns. Cambridge, UK: Cambridge University Press.

Cavanagh, J.E. (1993). Ocean energy systems. In: *Renewable Energy – Sources for Fuels and Electricity*, Ch. 12, eds. Johansson, T.B., Kelly, H., Reddy, A.K.N. & Williams, R.H., pp. 513–547. Washington, DC: Island Press.

DOE (US Department of Energy) (1992). *Ocean Energy Program Review, Programs in Utility Technologies, Fiscal Years 1990–91*. DOE/CH10093–99, DE91002176. Washington, DC: US Department of Energy.

DOE (1995). *Annex 1: Technology Profiles, Secretary of Energy Advisory Board*. Washington, DC: Task Force on Strategic Energy Research and Development, US Department of Energy.

ETSU (Energy Technology Support Unit) (1995). *An Assessment of Renewable Energy for the UK*. Harwell, UK: Department of Trade & Industry, UK Renewable Energy Programme.

Greer, L.S., Hubbars, H.M. & Bloyd, C.N. (1995). Renewable energy in Hawaii, Part II. In: *Advances in Solar Energy*, Vol. 10, Ch. 9, ed. Boer, K.W., pp. 455–515. Boulder, CO: American Solar Energy Society.

Sanders, M.M. (1991). Energy from the oceans. In: *The Energy Sourcebook – A Guide to Technology, Resources and Policy*, eds. Howes, R. & Fainberg, A., pp. 257–297. New York: American Institute of Physics.

Further reading

Avery, W.H. & Wu, C. (1994). *Ocean Thermal Energy Conversion*. Oxford: Oxford University Press.

Charlier, R.H. (1998). Re-invention or aggorniamento? Tidal power at 30 years. *Renewable and Sustainable Energy Reviews*, 1, 271–289.

McCormick, M.E. & Kim, T.I.C. (eds.) (1987). *Utilization of Ocean Waves – Waves to Energy Conversion*. New York: American Society of Civil Engineers.

Wick, G.L. & Schmitt, W.R. (eds.) (1981). *Harvesting Ocean Energy*. Paris: UNESCO Press.

8 Nuclear fusion

8.1 Introduction

Nuclear fusion is arguably *not* a sustainable energy source at least not in the full sense of sustainability we have adopted here. Meeting one of our criteria, fusion energy *is* superabundant and, therefore, would be inexhaustible. Indeed, it was once regarded as the ultimate superabundant, "backstop" energy resource (Nordhaus, 1979) that would place a cap on the inevitable price rise of all exhaustible fuels. While its superabundance, like that of solar energy, makes renewability irrelevant, fusion does not meet the test of recyclability, as we have defined it here (see the Introduction).

Recyclability, in the inclusive meaning we gave it, means that the technology itself produces little or no waste products or pollutants which must be dealt with to protect health and the environment (Beder,1994). Fusion technology has a variety of radioactive products and while these may be less hazardous than those of nuclear fission, nonetheless, they must be dealt with in operation and during radioactive waste disposal. In addition, nuclear fusion, at least in one of its forms (magnetic confinement), seems to lead to large-scale power plants and concentration of infrastructure, which are thought to be contrary to the broader concepts of sustainable economic and social progress (again, see the Introduction). For reasons such as these, fusion would not be included in most lists of future inexhaustible energy technologies by backers of decentralized, "independent-power" alternatives (Flavin & Lenssen, 1994a,b).

Regardless of these philosophic considerations, however, it should be clearly recognized that fusion has not yet been proven technically feasible for practical use. That is to say, there has not yet been a proof of principle that a net yield of energy can be obtained from a "controlled fusion" reaction (Cassedy & Grossman, 1990a). Nuclear-fusion power has been the subject of research since the late 1950s (Dean, 1981; Stacey, 1984; Furth, 1990). These efforts followed closely on the development of the hydrogen bomb, which itself was a sudden, massive nuclear fusion reaction triggered by a nuclear-fission (bomb-type) reaction. Fusion power research, by contrast, has attempted to achieve a *controlled* fusion reaction that can be contained in a power-producing reactor. The nuclear reactions themselves occur between the isotopes of hydrogen – the lightest elements – and then only when the nuclei can be forced sufficiently close together against the strong repelling force of their (ion) charges. The only physical conditions that have been demonstrated to achieve this proximity of nuclei are extremely high (solar) temperatures, when these nuclei collide with one another at extremely high speeds in thermal motion. Consequently, the reactions are termed **thermo-nuclear** reactions, which is the basis for both the bomb and the power reactors.

Despite the imagery of bomb and reactor, controlled-fusion reactors are accepted as being far less hazardous than nuclear-fission reactors. These fusion reactions are not chain reactions nor do they have long-lasting (radioactive) decay heat, both of which make for dangerous kinetics in fission reactions (Cassedy & Grossman, 1990b). In addition, the radio-active isotopes in fusion, such as tritium, are much shorter lived and give off less potent radiation than do the products of fission. The ultimate hope of the controlled-fusion researchers is to obtain the "clean reactions," such as that between deuterium and the helium-3 isotope, which produces neither tritium nor high-energy neutrons. Both of these reaction products are sources of dangerous radiation and occur in the fusion reaction between deuterium and tritium (D–T), which is the one that occurs at lower (still solar) temperatures and the one first hoped to be achieved. The neutrons, besides being deadly at the time of their emission from nuclear reactions, activate radioactivity in the reactor walls and surrounding structures.

Despite decades of research by the most able of investigators worldwide as discussed widely in journals such as *Science* and *Physics Today* (Anon., 1990, 1994c; Cordey, 1992, Furth, 1992), nuclear fusion has still not had a proof of principle. That is, fusion will not be ready for development into a commercial power-producing technology until it is demonstrated that more energy is released from the nuclear reactions than it requires to create them. The first milestone to achieve in the reactor experiments is the "breakeven point", where the nuclear energy released at least *equals* the energy input to create the high temperatures and densities required (Stacey, 1984; Epstein, 1991). Beyond that point, "gain" must be attained, where more energy is released than required on input; beyond that still is achieving a steady-state reactor with continuous energy output.

To date, the reactor experiments in the USA and abroad (Anon., 1994a,b) have produced significant amounts (megawatts, thermal) of re-leased nuclear energy but have still not reached the breakeven point. Even if breakeven and gain are demonstrated, it will have been achieved using D–T fusion, which requires a tritium input and a nuclear reaction that produces high-energy neutrons. The "clean" reactions occur only at much higher temperatures (around a billion degrees Celsius) and will be even more difficult to achieve experimentally than the "dirty" D–T reaction.

8.2   Magnetic confinement

Progress has been made in recent years in fusion research toward these goals in spite of daunting problems in the physics of fusion plasma (Stacey, 1984; OTA, 1987). The major research efforts have been on the "tokamak" type reactor, in which the hot plasma is confined to a donut-shaped (torroidal) ring by strong magnetic fields. The original means of treating the plasma, by inducing huge (tens of megamperes) currents in the ionized plasmas, was found to be insufficient to reach the high temperatures (hundred million degrees Celsius) required for D–T fusion. After many years of experimentation, two supplemental means of heating were found: injection of high-energy, neutral-particle beams and transmission of high-power radio waves into the plasma, either one of which could supply the several tens of megawatts of thermal power needed to reach fusion temperatures. In addition, difficult magneto–hydro dynamic instabilities that previously had defied taming were controllable in the tokamak torroids.

However, other unstable distortions and gyrations of the plasma are still possible problems depending on magnetic confining forces, gaseous pressures, and alike. These challenges, together with diffusion losses of particles from the plasma and radiative losses, all of which have stymied controlled fusion research to varying degrees from its inception, are indications of the uncertainties that lie ahead, just to achieve the proof of principle. The next generation of experimental reactors will be built and operated over a period extending into the 21st century, if research funding is provided (Anon., 1990, 1994c). The largest and most advanced planned is the joint international tokamak project called INTER, but even its technical feasibilty has been questioned (Anon., 1996a,b). This will be an international effort, with participation by European countries and the US (further USA reactor prototypes have not been funded).

If the proof of principle is made for magnetic-confinement fusion, the technology will most likely be developed in the form of the tokamak reactor. The physical scaling of such plasma-fusion reactions, as well as the economies of size required for all auxiliary equipment, guarantees that the resulting power plants will be of large (multigigawatt electric output) capacities. (The INTER reactor, as designed, would be 30 m (110 ft) in height and in diameter and will produce about 1 GW_{th} fusion power). If heavy investment were to be made in such large centralized plants, it would be a step toward large-scale power plants with concentration of generation and transmission facilities, which is regarded by advocates of decentralized technology as contrary to sustainable progress.

8.3   Inertial confinement

Another approach to achieving fusion in the laboratory has been inertial confinement (Epstein, 1991; Anon., 1993) in which a small (1 mm diameter) pellet of D–T fuel would be heated by laser beams to fusion temperatures so rapidly (about 10^{-10} second) that the reaction could take place before the nuclei can fly apart. The dynamics actually model (in miniature) the heating, compression, and fusion ignition of the core of a hydrogen bomb, with the laser beams serving the function of the fission trigger. This being the case, the major motivation (and past funding) of the research and for its possible revival has been from nuclear-weapons programs. This weapons connection has also shrouded much of the inertial program in secrecy. Nonetheless, the technique of using laser (or ion) beams has been thought by many researchers to be a more straightforward way to demonstrate breakeven and, ultimately, to achieve useful energy output.

An inertial-confinement reactor would operate by firing the fusion pellets one at a time, each time causing a miniature fusion explosion yielding about 10 MJ (0.01 MBTU) thermal energy from the fusion neutrons. If a single reactor could be made to ignite several pellets per second, then a single reactor would have thermal outputs of several tens of megawatts (thermal), which might supply the prime mover for electric generation in the 10 MW_e power range. It is evident that such reactors, taken individually or even in cluster, would supply relatively small power plants.

Inertial fusion, like the magnetic version, has faced daunting problems in physics. The physical conditions created within the D–T pellet to trigger fusion are far beyond those ever achieved before in the laboratory, with compressions approaching 1000 times liquid densities and temperatures in the range of one billion degrees Celsius. The critical problem – grappled with for decades – has been to couple more of the laser-light energy into the fusion core of the pellet, rather than scattering the light away or merely heating the free electrons in the plasma. A solution was proposed several years ago (Nuckolls, 1982; Anon., 1993) called "indirect drive", which couples the incident laser energy through a surrounding shell of a heavy metal called a "hohlraum". This shell better absorbs the light, and the heavy (metal) nuclei emit X-rays when heated to very high temperatures, thus helping to compress the D–T pellet. This technique was kept secret for years, however, because it was related to weapons design. It has recently been promoted, with great optimism as part of a newly proposed project to achieve breakeven with inertial fusion.

In 1979, an ad hoc review committee on inertial fusion for the DOE concluded its report: "we can see no insurmountable roadblocks to the achievement of electrical power generated by inertial-confinement fusion . . . " (Nuckolls, 1982). Fifteen years later, however, a breakeven reaction was still forecast for the future, and then only after the completion of a new facility costing over $500M. This new facility would employ approximately 200 lasers, which collectively would deliver 2 MJ light energy to the fusion target, less than 1 mm in diameter, in 10^{-10} seconds. With this much laser energy driving the tiny pellet, through the hohlraum, it is hoped that the fusion reaction will exceed breakeven. Even if attained, however, it should be recognized that this is a breakeven in principle only, since it is defined as the ratio of the reaction energy output to the energy of the incident laser light and, therefore, does not account for the energy efficiency of the lasers themselves. Since the lasers used have efficiencies in the range of 10%, this would require a fusion gain over 10:1 in order to attain an overall energy breakeven. The low efficiencies of the high-power lasers required to drive the pellet targets have not been overcome in decades of research, however.

8.4   Cold fusion

In 1989, startling news rang out that nuclear fusion had been achieved in an ordinary laboratory beaker at room temperatures (Close, 1991). The experiments, soon dubbed "cold fusion", attracted great attention because of the seeming possibility of being able to tap the superabundant resources of fusion without the extraordinary measures of solar-temperature plasmas just described. The claims in the media of the two Utah researchers to have observed cold fusion, however, were soon hotly disputed (Crease & Samios, 1989; Trower, 1989).

The theory upon which the supposed phenomenon was based rested on the ability of certain metals (e.g. titanium and palladium) to absorb large quantities of hydrogen (or its isotopes) into their interatomic crystalline structures. The packing of the isotope deuterium was thought possibly to be close enough, with the aid of quantum mechanical actions at the atomic scale, to allow nuclear fusion to occur. While these theories were never conclusively discredited, all experimental observations claiming evidence of nuclear processes were heavily disputed by other experimentalists, some of whom reported negative outcomes to their attempts to repeat the experiments of the cold-fusion advocates.

Assessments of the cold-fusion claims by these critics ranged from sloppy experimental techniques on the part of the advocates to charges of out-and-out dishonesty (Close, 1991). One analysis of the episode even likened the advocates' claims to earlier instances of "pathological science" (Langmuir (reprinted), 1989). Subsequently, some experiments that might be called "cold inertial confinement" have been pursued by other researchers (Browne, 1994), but strong interest in cold fusion has died out, as has most funding support for further research.

8.5   Summary

Nuclear fusion holds a long-term uncertain promise for its proposed role as an inexhaustible energy resource. As of this writing, a proof of principle for net energy gain has not been achieved. However, even if net energy gain is proven and a nuclear-fusion technology for power plants is developed, its likely evolved form might be incompatible with sustainable energy planning because of the large size of the reactors and the attendant problems of operating a nuclear technology.

References

Anon. (1990). Europe: betting heavily on fusion (News and Comment). *Science*, 250, 1500–1502.

Anon. (1993). Laser fusion catches fire (News and Comments). *Science*, 262, 1504–1506.

Anon. (1994a). Princeton Tokamak begins experiments with tritium–deuterium (News). *Physics Today,* Jan., 17–18.

Anon. (1994b). Magnetic fusion tops limit at Princeton (News and Comments). *Science,* 266, 1471.

Anon. (1994c). Fusion research at the Crossroads (News and Comments). *Science,* 264, 648–651.

Anon. (1996a). European Report Champions ITER. (Fusion Research). *Science,* 274, 1603.

Anon. (1996b). Turbulence may sink Titanic reactor (News and Comment). *Science,* 274, 1600–1602.

Browne, M.W. (1994). New shot at cold fusion by pumping sound waves into tiny bubbles. (Science Times). *New York Times,* Dec., C1, C10.

Beder, S. (1994). The role of technology in sustainable development. *IEEE Technology and Society Magazine,* 13, 14–19.

Cassedy, E.S. & Grossman, P.Z. (1990*). Introduction to Energy – Resources, Technology and Society.* Cambridge, UK: Cambridge University Press. (a) Nuclear fusion power reactors, Ch. 12, pp. 252–262; (b) Nuclear fission technology, Ch. 7, pp. 134–166.

Close, F. (1991). *Too Hot to Handle: The Race for Cold Fusion.* Princeton, NJ: Princeton University Press.

Cordey, J.G. (1992). Progress toward a tokamac fusion reactor. *Physics Today,* Jan., 22–30.

Crease, R.P. & Samios, N.P. (1989). Cold fusion confusion. *The New Yorker Magazine,* Sept., 34–38.

Dean, S.O. (1981). *Prospects for Fusion Power.* New York: Pergamon Press.

Epstein, G.L. (1991). Fusion technology for energy. In: *The Energy Source Book – A Guide to Technology Resources and Policy,* eds. Howes, R. & Fainberg, A. pp. 153–173. New York: American Institute of Physics.

Flavin, C. & Lenssen, N. (1994a). *Empowering the Future: Blueprint for a Sustainable Electricity Industry,* Paper 119. Washington, DC: World Watch Institute.

Flavin, C. & Lenssen, N. (1994b). *Power Surge – Guide to the Coming Energy Revolution.* The World Watch Environmental Alert Series. Washington, DC: Worldwatch Institute.

Furth, H.P. (1990). Magnetic confinement fusion. *Science,* 249, 1522–1527.

Langmuir, I. (1989). Pathological science (transcribed and reprinted, ed. Hall, R.N.). *Physics Today,* Oct., 36–48.

Nordhaus, W.D. (1979). *The Efficient Use of Energy Resources.* New Haven, CT: Yale University Press.

Nuckolls, J.H. (1982). The feasibility of internal-confinement fusion. *Physics Today,* Sept., 24–31.

OTA (Office of Technology Assessment) (1987). *Star power: the US and International Quest for Fusion Energy.* OTA-E-338. Washington, DC: Office of Technology Assessment, US Government Printing Office.

Stacey Jr, W.M. (1984). *Fusion – An Introduction to the Physics and Technology of Magnetic Confinement Fusion.* New York: Wiley-Interscience.

Trower, W.P. (1989). Cold fusion as seen by X-ray vision (Letter). *Physics Today,* July, 13.

Further reading

Holdren, J.P., Berwald, D.H., Budnitz, R.J. *et al.* (1988). Exploring the competitive potential of magnetic fusion energy: the interaction of economics with safety and environmental characteristics. *Fusion Technology*, 13, 7–56.

National Research Council (1986). *Plasmas and Fluids.* Washington, DC: Panel on the Physics of Plasmas and Fluids, National Academy Press.

9 Hydrogen fuel from renewable resources

9.1 Background

Hydrogen has been promoted as the universal fuel for the future, with all of society's energy needs supplied in a "hydrogen economy" (Bockris, 1975; Dickinson *et. al.*, 1977; Ogden & Williams, 1989). Its advocates emphasize that it is clean burning and, as a resource, is inexhaustible.

Hydrogen is indeed a nearly ideal fuel, since it produces little pollution and emits no greenhouse gasses when burned. Although combustion of hydrogen itself has only water as the product of combustion, this will only be the observed product if pure oxygen is used for the process. If hydrogen is burnt in air, which is 78% nitrogen and 21% oxygen, the nitrogen will also be oxidized to give various nitrogen oxides, just as occurs when fossil fuels are burnt in air. Nonetheless, other pollutants, such as sulfur dioxide, are not emitted nor is carbon dioxide (greenhouse gas). No nitrogen oxides or other noxious pollutants are emitted from the reactions of hydrogen in certain types of fuel cell.

It should be recognized at the outset, however, that hydrogen is *not* a primary energy resource. It is not found in a free gaseous form in nature, as methane is. It must always be dissociated from a naturally occurring compound, such as water (H_2O) or natural gas (methane, CH_4). For this reason, hydrogen has been called an energy "vector" or carrier (Winter & Nitsch, 1988; Imarisio & Strub, 1983), meaning that it serves as a means of transmitting and distributing energy but is itself not the primary source of the energy. A primary energy source is always required to accomplish dissociation to free hydrogen. For our purposes here, we will consider only water as the resource, since natural gas (a fossil fuel) would not fit into the category of a sustainable resource as defined here (see the Introduction).

9.2 History

Notions of using hydrogen as a fuel have an interesting history, revealing a great deal about the present-day research effort. Hydrogen has held a fascination for scientists and pseudo-scientific thinkers going back to its discovery in the late 18th century by Lavoisier and Cavendish (Hoffman, 1981). Its lighter-than-air properties were quickly recognized and hydrogen balloons were being flown shortly after the French Revolution. Water was dissociated by an electric current in 1818 in England shortly after Volta built his first electrolytic cell. Its use to power machinery was first proposed in 1820 in a treatise by William Cecil at Cambridge University and 9 years later the first fuel cell (as we would now know it) was proposed.

Further scientific work on hydrogen did not occur until the early 20th century. However, a work of science fiction was to capture the imagination of many, including later scientists. In 1874, Jules Verne wrote *The Mysterious Island*, in which the major characters find themselves stranded on an island where the inhabitants know how to use water as a fuel. Verne did not explain, and most probably did not know, that it required an input of energy to dissociate water in order to create such a fuel. Nevertheless, the idea of an unlimited and ubiquitous resource enthralled readers for decades thereafter. One of the pioneers for hydrogen-powered automobiles in the early 19th century, F. Lawaczek (German), is said to have been inspired by Vernes' book.

The first modern research on hydrogen fuels was carried out in Germany and the UK during the 1920s (Hoffman, 1981). In a 1923 lecture at Cambridge, J. B. S. Haldane proposed wind power as the source of electricity to electrolyze water – over a half century ahead of its realization. In 1928, Rudolph Erren, a pioneer of hydrogen technology in Germany, obtained his first patent for a hydrogen engine. Prototype hydrogen-propelled rail cars were built in Germany in the late 1930s. Interest in hydrogen propulsion in Germany was paralleled by its use in zeppelins in the 1930s. Attempts were made during World War II to achieve hydrogen aircraft engines in order to replace scarce petroleum fuels with fuels derived from coal. This motivation to obtain coal-derived fuels also led to a more extensive use of hydrogen in German industry than in other industrial countries.

Following the war, some interest developed in hydrogen as a means of transmitting energy; this concept was later to be called an "energy carrier" (Winter & Nitsch, 1988) or "energy vector" (Imarisio & Strub, 1983). These ideas arose from the newly emerging nuclear-power interests in the 1950s. An outstanding advocate of hydrogen-energy carriers was Cesare Marchetti of the EURATOM research center in Ispra, Italy, who proposed that the energy output of nuclear reactors be delivered either as electricity or as a hydrogen fuel. He pointed out that energy in the form of hydrogen can be stored more readily than electricity (see Chapter 5). Marchetti also noted that transmission costs (per unit of heat energy delivered) would be less for hydrogen than for electricity, a fact verified a few years later with engineering data (Hottel & Howard, 1971).

Frequently, in this evolution of hydrogen-fuel concepts, grandiose schemes have been proposed (see Hoffman (1981) for a review). For instance, in 1938 Igor Sikorski prophesied that 400 years hence, Great Britain would supply all its energy needs by "rows of metallic wind mills working electric motors (sic) which supply current . . . for electrolytic decomposition of water". Further in this vein, Marchetti proposed in the 1950s that huge nuclear energy islands in the Pacific ocean could supply massive amounts of electricity and hydrogen, feeding major loads on shore.

By 1970, at the General Motors Technical Center, the concept of the "hydrogen economy" was advanced in which all the energy delivered to the various sectors of the economy would be in the form of hydrogen and electricity. The primary energy sources would be in massive nuclear power plants (nuclear fusion, if it became feasible) generating electricity and electrolyzing water (Hammond *et. al.*, 1973). The hydrogen economy was the subject of an intensive feasibility study by the Stanford Research Institute in 1976 and was reported in the scientific literature as a serious prospect during that period (Gregory, 1973; Goltsova *et. al.*, 1990; Dickinson *et. al.*, 1997). Hydrogen proponent Bockris prophesied in 1977 that: " . . . the hydrogen economy could be well on its way by 1990".

In the early 1970s, spurred by the oil crisis, enthusiasts for hydrogen technologies organized and formed the International Association for Hydrogen Energy (IAHE). The IAHE has subsequently instituted a research journal (the *International Journal of Hydrogen Energy*) and a series of research conferences: World Hydrogen Energy Conferences (e.g. the 8th World Hydrogen Energy Conference (Veziroglu & Takahashi, 1990)). Both efforts report on research in recognized research areas, for example solar energy, solid-state physics, electrochemistry, biochemistry and thermochemistry. They do so with much the same air of scientific objectivity as other professional societies. Still, at their world conferences and in their proceedings there appear advocacy statements promoting hydrogen technology that are unusual for such a professional society. For example, in the Foreword to the 8th World Hydrogen Energy Conference, the IAHE announces: ". . . our goal of establishing hydrogen as the fuel which powers the world economy". In the same proceedings volume, a resolution on "energy and environment" is printed, as signed by many of the conferees, to "ensure that eventually the present fossil fuel system . . . be replaced by . . . the Hydrogen Energy System". The IAHE, in short, appears to claim that a universal fuel system of hydrogen is *the* solution to the world's environmental and resource problems.

9.3   Hydrogen production

The most straightforward means to dissociate hydrogen is by electrolysis of water (Winter & Nitsch, 1988). Electrolysis is accomplished by inserting electrodes in a solution of water, to which an electrolyte chemical has been added (Fig. 9.1). When a d.c. voltage is applied across the electrodes, hydrogen evolves from the solution near the cathode and oxygen from the anode region. This comes about from the creation of ions at one electrode and the migration of those ions to the other electrode, which is of opposite polarity. When the ions reach the electrode, they recombine to complete the reaction.

Figure 9.1 shows two possibilities for electrolytes, alkaline or acid, having negative hydroxyl (OH^-) or positive hydrogen ions (H^+)), respectively. Electrolyzers have been in commercial operation since early in this century, in Europe and other industrial countries, to supply hydrogen for industrial processes. Development programs are underway for electrolyzers to increase their efficiencies and lower fixed and variable costs

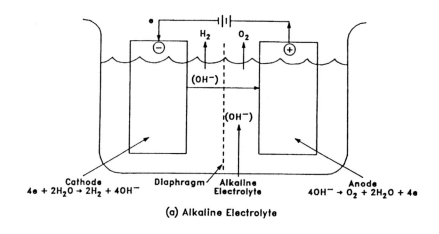

(a) Alkaline Electrolyte

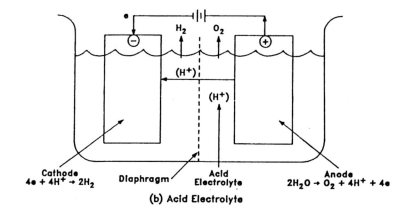

Figure 9.1
Hydrogen hydrolysis
cells: (a) alkaline
electrolyte; (b) acid
electrolyte. (From Boer
(1989), by permission
of the American Solar
Energy Society.)

(b) Acid Electrolyte

(Crawford & Benzimra, 1986). Figure 9.2 shows an operating prototype installation in Quebec of advanced electrolyzer cells.

In recent years, advanced electrolyzers have been developed incorporating more durable electrodes, improved circulation of electrolytes, and better separation of the output gasses (hydrogen and oxygen). Alternatively, the cells can be operated at elevated temperatures, with water vapor being electrolyzed, giving improved conversion efficiencies. Advanced electrolyzer cells have achieved energy conversion efficiencies greater than 80%. Currently, hydrogen fuel can be produced at unit costs less than $10/MBTU ($9.50/GJ) only if electricity costs are less than $0.02/kW-hr (rarely achieved). With natural gas prices being less than half that of electrolytic hydrogen, the latter is uncompetitive, even with the lowest existing electric rates.

A primary source of energy is required to supply the electricity for electrolysis. If this source is conventional electricity, generated from fossil fuels, the environmental benefits of hydrogen are largely lost. If electricity sources are used that are not polluting and do not emit greenhouse gases, for example solar, wind or hydro, the hydrogen-fuel system becomes environmentally benign from start to finish. The major emphasis of

Figure 9.2
Advanced electrolysis
cells in operation.
(Courtesy of IREC
(Hydro Quebec).)

present-day advocates of such hydrogen systems has been on solar–electric
sources and consequently the promotional term *solar–hydrogen* has been
adopted (Bockris *et al.*, 1989; Ogden, 1990; Winter, 1990).

The most straightforward route to solar hydrogen is using PV cells to
supply the electrolyzer cells. Both technologies are functional. The development of PV cells is discussed in Chapter 1 (p. 47) and these are acknowledged to be a working, albeit expensive, technology. Electrolyzers are also
working, but in need of cost reductions. Nonetheless, a prototype solar–
hydrogen plant has been designed for operation in Saudi Arabia (Winter &
Fuchs, 1991; Abaoud & Steeb, 1998).

9.4   The costs of solar–hydrogen

The principal obstacle to adoption of solar–hydrogen fuels is the cost of
PV-generated electricity. The production cost of electrolyzed hydrogen
will depend directly on the cost of the electricity used, no matter what the
electric source. In the PV case, we have seen unit costs exceeding $0.25/
kW-hr, based on unit (system) capital costs over $4/$W_p$. This would lead to
production costs (per unit of heating value) for hydrogen over $75/MBTU
($71/GJ), which is about 15 times the current price of natural gas. If,
however, optimistic cost reductions (Ogden, 1990) for PV come true,
giving electricity costs under $0.10/kW-hr (capital costs of $1.2/$W_p$), then
the production cost for solar hydrogen would drop to about $30/MBTU
($28/GJ). The long-term goal of the DOE Photovoltaics Program is
$0.065/kW-hr, which if achieved would yield a projected production cost

of about \$23/MBTU (\$22/GJ) for hydrogen (still over four times the current market price of natural gas).

These unit heating costs for a fuel are, of course, far from being economic against competing fossil fuels. With the present electrolyzing technology, furthermore, only a zero cost of electricity (Ogden & Williams, 1989) would bring hydrogen within the current competitive (\$2–3/MBTU or \$1.90–2.84/GJ) range of natural gas (at the well head). While more advanced electrolyzers are being researched, including the high-temperature scheme mentioned above, projections have been made only for modest reductions in the capital costs of the new electrolyzers over the next 5–15 years. Conversion efficiency improvements of 10–15% do appear to be possible with the advanced electrolyzers, leading to incremental reductions within this same percentage range for the cost of hydrogen. The major change in electrolyzer costs must come in capital costs, however, in order to reduce the total (solar electricity plus electrolyzer) costs to a competitive range.

Advocates for solar–hydrogen, however, expect technology improvements in PV cells exceeding DOE goals (Plass *et al.*, 1990), claiming possible cost credits for electrolyzers through sales of oxygen. The most extremely optimistic projection (Ogden & Williams, 1989) shows PV electricity costing \$0.02/kW-hr for an 18% PV-cell efficiency operating in the sunbelt, with a capital cost of \$0.2/W_p. This results in hydrogen production costs around \$9/MBTU (\$8.50/GJ), which despite its incredibly low projection for PV costs still yields a hydrogen cost about three times the present market price of natural gas.

9.5   Other renewable sources for hydrogen

Other sources of electricity are also possible, of course. Solar–thermal generation (p. 32) is a candidate technology. The cost of solar-thermal electricity, as achieved in prototype operation, is about \$0.08/kW-hr (Becker, 1992), which would result in a hydrogen production cost around \$28/MBTU (\$26/GJ) if we followed the cost formulae cited above for PV generation. It should be recognized for either form of solar generation, moreover, that the best prospects for economically competitive performance will only be found in the sunbelt regions, where the most successful demonstrations of both PV and solar-thermal projects have been operated. In other regions, the northeastern USA for example, the annual solar energy collected can be expected to be less than half that of the sunbelt, thus more than doubling the unit cost of solar electricity. We are forced to conclude that the solar-hydrogen option, if it is to be proven economic at all, will likely only be feasible in the sunbelt. It would, therefore, appear that either the use of hydrogen fuel would have to be restricted to those regions, or the hydrogen would have to be transmitted long distances to other regions, as natural gas presently is.

Ogden (1993) and Bockris and co-workers (1989) have suggested that hydrogen can feasibly be transmitted over long distances. They project that hydrogen can be transmitted through large-diameter pipes (5 ft, 1.5 m), specially designed to avoid hydrogen embrittlement of metals and with special provisions for the compression and flow characteristics of hydrogen. Such pipelines would cost 50% more than natural gas lines for the same distance and same energy transmitted because of these special technical requirements. For a 1000 mile (1600 km) hydrogen pipeline, for instance, the cost is projected to be at least $2/MBTU transmitted. This added to the most optimistic production cost of $9/MBTU would bring the projected delivered cost (not yet the selling price) to $11/MBTU, which, again, is much higher than current natural gas prices (delivered). Other estimates of the cost of delivered hydrogen, including regional transmission and local distribution, show a cost range of $14–35/MBTU ($13–33/GJ) (Plass, *et al.*, 1990). If storage is required, then costs of over $3/MBTU ($2.80/GJ) must also be added.

Another way to judge the projected costs of solar hydrogen in the sunbelt region is to compare the PV-generated electrolyzing costs at the (sunbelt) site with well-head prices of natural gas, based on wells located in the sunbelt. In the mid 1990s, well-head prices for natural gas from the US southwestern states were less than $2.50/MBTU ($2.38/GJ), with low-demand prices around $1.65/MBTU ($1.56/GJ) (Anon., 1992).

Other energy sources – such as the renewables wind, hydro, and biomass – are also possible routes to hydrogen fuels (Mathusa, 1979). Wind and hydro provide alternative non-fossil and non-nuclear means of generating electricity for electrolysis. Biomass provides alternative naturally occurring organic compounds from which hydrogen can be derived. All three have regional limitations for large-scale development, just as PV generation does. For example, wind-energy regional characteristics defining economic operation have been reviewed in Chapter 3. Hydro power plant siting is also restricted by regional water-shed patterns and land-use barriers, as reviewed in Chapter 4. Finally, biomass energy crops have regional restrictions, as we have seen in Chapter 2 , for significant expansion on a national scale.

In terms of presently achieved electricity costs, wind generation is more economic than photovoltaics to supply hydrogen electrolysis. It is in the cost range $0.07–0.09/kW-hr, as experienced from operating projects of significant sizes. Wind, like solar, is constrained regionally for optimally economic annual production. In addition, wind-generation capability is sensitive to local siting in many cases, as our example of the Altamont Pass showed. Restrictions on land use will exclude more prime sites for wind than for solar (Grubb & Meyer, 1993). Much of the available land is in the Upper Midwest (Brower *et al.*, 1993; Wager, 1994); consequently this region could take on the same role for wind as the southwest does for solar for the mass production of hydrogen. Even with the long-range goal of $0.04–0.05/kW-hr (DOE Wind Program), however, the delivered cost of hydrogen to a market 1000 miles away would be in the range $17–21/MBTU ($16–20/GJ) using the same calculation method as used here for PV .

A wind–hydrogen project is under way in Hawaii under the joint support of the DOE, the state of Hawaii, and several private sector organizations (Neill, *et al.*, 1990). The objectives in this project are limited to local use (on each Hawaiian island), starting with electrolyzed hydrogen as a storage medium that is later used as a domestic fuel. The islands have several locations with high average wind velocities, owing to the prevailing winds and the island terrains. The hydrogen storage will be used in leveling the intermittency of wind generation by feeding fuel cells to supply electricity when wind velocities falter.

Since the Hawaiian islands are isolated from any outside power system interconnections and none of the islands are interconnected electrically, this situation fits the category of "remote operation" discussed in Chapter 1. As we learned there, service operability for remote sites makes for its own cost justifications. Therefore, wind generation, which is marginally economic for interconnected electric grid service, shows possibilities of benefit even with the added cost of hydrogen conversion and storage. What is achieved is uninterrupted electricity supply in stand-alone operation, for which there may be a willingness to pay the extra cost. In addition, there are the motivations to relieve the islands from dependence on oil imports and to move to non-polluting renewable energy sources. It should be recognized, however, that such a geographical situation is not common.

Hydro power offers some of the lowest cost prospects for renewable, non-polluting means to generate electricity. It could, therefore, be argued that hydro-generated electricity is the best sustainable means to electrolyze water, as indeed it is in the province of Quebec with extensive hydro resources. However, the costs of hydro power delivered to its end user are significantly dependent on transmission costs, since feasible hydro project sites often can be found only in remote locations (Chapter 4). Consequently, since these factors and financing costs can vary, hydro energy costs range from under $0.025/kW-hr to several cents per kilowatt-hour (Moreira & Poole, 1993). At the lower end, of course, hydroelectric-driven electrolyzers could produce hydrogen at about $5/MBTU ($4.70/GJ), or about half the cost of most optimistic PV solar–hydrogen projections. In fact, pilot demonstrations have been operated at high efficiencies and relatively low costs (Mathusa, 1979). It should be recognized, however, that hydroelectric power is a renewable resource in its own right. There may be no compelling reason to take what may be the lowest cost electricity available to the market and divert it to hydrogen production, unless it is part of a broader scheme of hydrogen use. Hydro power is presently utilized over its range of costs according to its economic competitiveness in the regions where it is generated.

The concept of a universal energy system could be used to argue for hydrogen (pipeline) transmission as a means to deliver hydroelectric power to remote loads instead of using electric transmission. Hydrogen advocates assert that long-distance energy transmission via hydrogen pipelines is far less expensive than by electric transmission lines (Ogden & Williams, 1989). (The correctness of their assertions would depend on the technology choices in electric transmission lines (Wu, 1990) and for pipeline systems (Hottel & Howard, 1971).) At the receiving end of the pipeline, the hydrogen would either be used as a fuel or converted back to electricity (e.g. using hydrogen fuel cells). In any such scheme, however, the price of the energy delivered would have to reflect the generation costs at the source. Hydrogen transmission on a large scale could be considered as an alternative to multigigawatt electric transmission systems for the massively ambitious plans for large hydroelectric projects in Africa and South America (IEEE, 1993). Such large-scale transmissions of energy in the form of a hydrogen carrier could fit with the mass scale of the contemplated investment in Third World hydro projects (see p. 144). Again, such proposals would need to be assessed against the realities of cost effectiveness and environmental impact concerns.

Returning now to hydrogen conversion, it is also possible to use biomass feedstocks as sources of hydrogen. In Chapter 2, we learned of the thermochemical means of producing combustible gasses, such as hydrogen and methane, from organic matter. The approach favored in the current research programs for biomass gaseous fuels (Miles & Miles,1989) is to perfect methanation processes in order to increase the proportion of methane in the thermogasification mixtures. Some have even proposed solar–thermal sources to obtain the required heat (Rustamov *et al.*, 1998). The objective in that R&D program is to attain nearly pure methane as a "pipeline-quality" gas substitute for (fossil) natural gas. An alternative to increased methanation, however, is to increase the proportion of hydrogen in the biogas.

A steam-reforming reaction (DOE, 1992; Ogden, 1993; Shiga *et al.*, 1998) on the biogas mixtures coming out of the initial thermochemical gasification stage will convert the mixture to a higher fraction hydrogen, at the expense of the methane. The final product of a two-stage process is a mixture of hydrogen and carbon dioxide, from which the CO_2 can be separated to leave nearly pure hydrogen. (Steam-reforming of hydrocarbons, such as natural gas and petroleum fractions, is a known method of producing hydrogen.) It is interesting to note in passing that pilot-scale, methane-reforming processing was carried out in the former USSR, using a parabolic solar furnace to supply the required heat (Parmon, 1990). While this biomass-derived hydrogen scheme appears technically feasible, little development work has been done to create a working technology. This also means, of course, that the production costs of such processing are unknown. Conceptually, a biomass-based hydrogen technology is no less desirable from the standpoint of sustainability than the solar-hydrogen technology emphasized by its proponents. As we learned in our section on biomass fuels, the carbon cycle can feasibly be closed with energy crops, thus ideally eliminating net emissions of CO_2. Also, the biomass resource is renewable, even in the strict sense of the term, if appropriate techniques are used for energy crops, as described in Chapter 2.

Recommendations have been made in more recent years by some hydrogen advocates to give a higher priority for biomass hydrogen R&D (Ogden & Nitsch, 1993). Still other long-range prospects for hydrogen production lie in photochemical processing, including photoelectrochemical and photobiochemical reactions (Winter & Nitsch, 1988; Parmon, 1990;

Bolton, 1996). Perhaps the most straightforward connection to the main-line concepts in hydrogen technology is photoelectrochemical conversion, in which water is dissociated using sunlight in one step without a separate source of electricity. This is accomplished using a photovoltaic cell combining (solid) semiconductors with a liquid electrolyte (Murphy & Bockris, 1983; Ohta, 1988; Khaselev & Turner, 1998) at each of two electrodes. The effect uses an aqueous acid electrolyte together with catalyst surfaces at the solid–liquid interfaces. It was pioneered by Fujishima and Honda (1972) to create the equivalent of a p–n junction region at the solid–liquid interfaces, with an electric potential that is driven by the incident solar energy. In addition, charge carriers are created (again similarly to the purely solid PV cell) that combine with ions in the electrolyte to create hydrogen and oxygen; thus an electrolyzer is created that does not require an external source of electricity.

These solar-driven techniques have been the subject of research investigations for over two decades. Further work has been done by Bockris and others (Bockris *et al.*, 1989; Bockris & Gonzalez, 1990; Parmon, 1990). Low conversion efficiencies (less than 10%) and decomposition of the catalytic electrodes have been found in the course of research, both of which can be addressed by searching for new catalyst compounds for coating the electrodes. This research effort, taking place at laboratories scattered around the globe, illustrates the speculative nature of new concepts in these technologies. While the promise of dramatic innovations remains, the timetable for any one of them to come to fruition is highly uncertain (see Chapter 11).

Other avenues of research are also being pursued, including various thermochemical and biochemical processes. The thermochemical reactions, in addition to ones operating on biomass, are potential alternatives to electrolysis of water or are means of deriving hydrogen from other compounds (Veziroglu & Takahashi, 1990). Some propose solar–thermal systems as a means of providing the heat required for the reactions. Photobiological reactions to produce hydrogen are also interesting alternatives (Veziroglu & Takahashi, 1990; NREL, 1992). Several different schemes utilizing bacterial action have been proposed and tried on a laboratory scale. These include the use of bacteria or algae that are otherwise nitrogen fixing but, with the proper enzymes, can be made hydrogen evolving. Fermentation processes involving anaerobic bacteria have also been studied in the laboratory, as well as so-called "phototropic" bacteria that can feed on starch and cellulosic wastes.

9.6   Hydrogen end uses

Hydrogen can either be used as a fuel in combustion or it can used in a fuel cell to generate electricity or to power an electric vehicle (Hoffman, 1981; MacKenzie, 1994). When burned as a fuel in air, the nitrogen oxides (products of combustion) can reportedly be reduced by using inexpensive catalytic converters (Ogden & Nitsch, 1993), similar in action to the

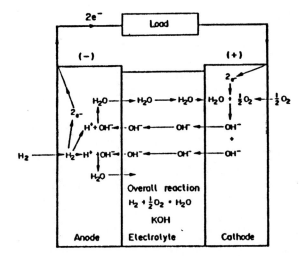

Figure 9.3
A hydrogen/oxygen fuel cell with porous electrodes, (From Kordesch & Oliveira, *International Journal of Hydrogen Energy*, Vol. 13, No. 7, 1988; reprinted by permission of the International Association for Hydrogen Energy.)

catalytic converters used on gasoline-fueled automobiles. The hydrogen fuel can be used for either stationary applications, in place of natural gas, or in automotive ICEs in place of petroleum fuels (see Chapter 5). The use of hydrogen in a fuel cell, however, provides an alternative means of generating electricity and an alternative form of transport: the FCV (see Chapter 5).

Fuel cells supply electricity through chemical reactions, as do batteries, but require a constant input of the reactants to keep the process going. Figure 9.3 is a schematic of a hydrogen fuel cell where the reactants are hydrogen fuel and oxygen (from the air) (Kordesch & Oliveira, 1988). The chemical reaction in this particular (alkaline electrolyte) cell is between hydrogen and hydroxyl ions, causing an exchange of charges for the flow of electricity. Fuel cells generate d.c. electricity, which means that, like solar PV cells, inverters are needed if they are to feed a.c. electric grids. In automotive applications using d.c. traction motors, the cell output can be used directly.

Whereas other fuels can be used in fuel cells, hydrogen provides a reaction that is virtually pollution free, which makes a FCV vehicle (Lemons, 1990; Ascoli, 1991; Billings *et al.*, 1991) using hydrogen a candidate zero-emission vehicle (ZEV). Low emissions are also an advantage in stationary applications, such as dispersed generation schemes. These benefits might tend to promote a willingness on the part of the public to pay the cost penalties to produce this fuel or the extra cost burdens (Chapter 5) for hydrogen storage. Safety is still a concern with hydrogen (De Luchi, 1989), as the hazards of explosion are greater than with natural gas or gasoline, especially in confined spaces.

9.7   Summary

In closing, it should be recognized that hydrogen, as a universal fuel and energy carrier, is a tantalizing prospect. Were it to replace oil and gas for these purposes, it could substitute for hundreds of quads of primary energy annually worldwide, costing tens of trillions of dollars each year. It is, however, a long-range and highly uncertain prospect. This seems to be the reality despite the optimistic forecasts of its promoters. Hydrogen production technology, by means of PV–electrolysis (solar–hydrogen) or biomass conversion, is still very much in the development stage. Prototype operations have yet to achieve competitive production costs, and the prospects for doing this in the medium term are not good unless the most optimistic forecasts of the promoters are realised. These projections are, if anything, even more optimistic for achievement of timescale goals than those of the proponents of PV (Chapter 1). Other sustainable means of hydrogen production, such as photochemical or biochemical methods are still in the stage of laboratory research and so projections of timely achievement for these concepts are hardly possible.

The DOE takes a tempered and limited view of the prospects for hydrogen technologies (DOE, 1992). It looks for "practical pathways" to applications in transportation, electric utilities and industry and sees prospects for "multi quad" use for the USA in the medium to long term only. (By multi quad they appear to mean several but less than 10 QUAD.) Total US primary energy demands in the medium term should be approximately 85 QUAD (or 85 EJ) in the USA (present US consumption is about 85 QUAD). The timescale for these DOE objectives is linked to the expressed need for the USA to become independent of fossil fuels by the end of the 21st century. The DOE sees these applications as economic only after more economic production technologies have been developed and views direct solar (not PV–solar) or biomass conversion as the most promising candidates for low-cost production in the long run.

The DOE Hydrogen Research Program is funded largely under the *Matsunaga Hydrogen Act* of 1990, named after the late Senator Spark M. Matsunaga from Hawaii, who was its principal sponsor in the Congress. The act directed the Secretary of Energy to prepare a 5-year plan to identify economic hydrogen technologies that are fed from renewable primary sources. Under the act, $7M was appropriated for fiscal year 1993 and $10M was projected for fiscal year 1994. The total DOE funding for hydrogen-related research and development in fiscal year 1992 was $81M (out of a total DOE budget of about $18B). This hydrogen funding included over $60M specifically for fuel cells. Support expressly for hydrogen sources was expected to increase for fiscal year 1994 and beyond, but only incrementally.

The first priority of the Matsunaga bill funding was for hydrogen-storage research, with a 45% allocation, followed by hydrogen-production investigations at a 35% allocation, thus suggesting an incremental approach to hydrogen production rather than attempts to achieve dramatic breakthroughs. Expenditures in other industrial countries are comparable, with Germany allocating $50M and Japan $20M in fiscal year 1989 (when the US funding was $3M) (DOE, 1992). The annual funding in Germany is expected to continue at least at this level for another decade from federal sources, supplemented by foundation funding of a solar and hydrogen research center (Winter, 1990). The German effort can be expected to be strongest because of the long history of hydrogen use there.

The conclusion one must draw from the level of support and limited objectives of these programs is that its expectations are nowhere near those of the "solar-hydrogen" community, for better or for worse. This view of limited prospects, and occurring in the long term only, is shared by at least some other advocates of renewable technologies (Brower, 1992). In short, the view of hydrogen as "the universal fuel powering the world economy," is not shared widely outside the circle of solar-hydrogen promoters. Nor are there projections of likely large-scale market displacement of fossil fuels by hydrogen within the foreseeable future emanating from any major groups outside that circle.

References

Abaoud, H. & Steeb, H. (1998). The German–Saudi HYSOLAR Program. *International Journal of Hydrogen Energy*, **23**, 445–449.

Anon. (1992). Wellhead prices *Oil & Gas Journal*, 90, Oct., 38.

Ascoli, A. (1991). Fuel-cell powered electric vehicles: overview and perspectives. *International Journal of Global Energy Issues*, **3**, 217–220.

Becker, N.D. (1992). The demise of LUZ: a case study. *Solar Today*, Jan./Feb., 24–26.

Billings, R.E. *et al.* (1991). Laser cell prototype vehicle. *International Journal of Hydrogen Energy*, **16**, 829–837.

Bockris, J.O'M. (1975). *Energy: The Solar–Hydrogen Alternative*. New York: Wiley.

Bockris, J. O'M. & Gonzalez, A. (1990). The photoproduction of hydrogen. In: *Proceedings of the 8th World Hydrogen Energy Conference on Hydrogen Energy Progress*, Hawaii, eds. Veziroglu, T.N. & Takahashi, P.K. pp. 791–800. New York: Pergamon Press.

Bockris, J.O'M., Dandapani, B. & Wass, J.C. (1989). A solar hydrogen energy system. In: *Advances in Solar Energy*, Ch. 3, ed. Boer, K.W., pp. 171–305. New York: Pergamon Press.

Boer, K.W. (1989). In: *Advances in Solar Energy*, Vol. 5, eds. Boer, K.W. & Duffie, J.A., p. 255. Boulder, CO: American Solar Energy Society.

Bolton, J.R. (1996). Solar production of hydrogen: a review. *Solar Energy*, **57**, 37–50.

Brower, M. (1992). *Cool Energy – Renewable Solutions to Environmental Problems*, revised edn. Cambridge, MA: MIT Press.

Brower, M.C., Tennis, M.W., Denzler, E.W. & Kaplan, M.M. (1993). Powering the Mid West – renewable electricity for the economy and the environment. In: *Biomass Energy*, Ch. 3. Cambridge, MA: Union of Concerned. Scientists.

Crawford, G.A. & Benzimra, S. (1986). Advances in water electrolyzers. *International Journal of Hydrogen Energy*, **11**, 691–701.

De Luchi, M.A. (1989). Hydrogen vehicles: an evaluation of fuel storage, performance, safety, environmental impact and costs. *International Journal of Hydrogen Energy*, **14**, 81–130.

Dickinson, E.M., Ryan, J.W. & Smulyan, M.H. (1977). *The Hydrogen Energy Economy – A Realistic Appraisal of Prospects and Impacts.* New York: Praeger.

DOE (US Department of Energy) (1992). *Hydrogen Program Plan.* FY 1993–FY 1997. DOE/CH 10093-147. Golden, CO: Office of Conservation & Renewable Energy, National Renewable Energy Laboratory.

Fujishima, A. & Honda, K. (1972). Electrochemical photolysis of water at a semiconductor electrode. *Nature*, 238, 37–38.

Goltsova, L.F., Garkusheva, V.A., Alimova, R.F. & Goltsov, V.A. (1990). Hydrogen energy and technology in the world (a scientometric study of the literature). *International Journal of Hydrogen Energy*, 15, 9, 655–660.

Gregory, D.P. (1973). The hydrogen economy. *Scientific American*, Jan., 13–21.

Grubb, M.J. & Meyer, N.E. (1993). Wind energy: resources, systems and regional strategies. In: *Renewable Energy – Sources for Fuels and Electricity*, Ch. 4, eds. Johansson, T.B., Kelly, H., Reddy, A.K.N. & Williams, R.H., pp. 157–212, Washington, DC: Island Press.

Hammond, A.L., Metz, W.D. & Maugh II, T.H. (1973). *Energy and the Future.* Washington, DC: American Association for the Advancement of Science.

Hoffman, P. (1981). *The Forever Fuel, The Story of Hydrogen*, Ch. 3: Early visions: the history of the hydrogen movement. Boulder, CO: Westview Press.

Hottel, H.C. & Howard, J.B. (1971). *New Energy Technology – Some Facts and Assessments.* Cambridge, MA: MIT Press.

IEEE (IEEE Power Engineering Review) (1993). *Power Engineering Society (PES) Winter Meeting Panel Session on International Electric Network History and Future Perspectives of the World Bank and the United Nations, Global Interconnections 1993*, pp. 12–24. (a) The World Bank's role in the electric power sector (R. Vedavalli, The World Bank, Washington, DC); (b) The United Nations technical assistance for developing countries (Y. Abu–Alam, United Nations, NY); (c) Electric power projects in Latin America (N. de Franco, The World Bank, Washington, DC); (d) Benefits and pitfalls of international connections (T.S. Drolet & J.S. McConnack, Ontario Hydro International, Canada).

Imarisio, C. & Strub, A.S. (eds.) (1983). *Proceedings of the 3rd International Seminar: Hydrogen as an Energy Carrier*, Lyon, France. Dordrecht, the Netherlands: Reidel.

Khaselev, O. & Turner, J.A. (1998). A monolithic photovoltaic–photoelectrochemical device for hydrogen production. *Science*, **280**, 425–427.

Kordesch, K. & Oliveira, J.C.T. (1988). Fuel cells: the present state of the technology and future applications with special consideration of the alkaline hydrogen/oxygen (air) systems. *International Journal of Hydrogen Energy*, 13, 411–427.

Lemons, R.A. (1990). Fuel cells for transportation. *Journal of Power Sources*, **29**, 251–264.

MacKenzie, J.J. (1994). *The Keys to the Car – Electric and Hydrogen Vehicles for the 21st Century.* Washington, DC: World Resources Institute

Mathusa, P.D. (1979). *Hydropowered electrolysis in New York State.* New York: Wiley.

Miles, T.R. & Miles, T.R. Jr (1989). Overview of biomass gasification in the USA. *Biomass*, 18, 163–168.

Moreira, J.R. & Poole, A.G. (1993). Hydropower and its constraints. In:

Renewable Energy – Sources for Fuels and Electricity, eds. Johansson, T.B., Kelly, H., Reddy, A.K.N. & Williams, R.H., pp. 73–119. Washington, DC: Island Press.

Murphy, O.J. & Bockris, J.O'M. (1983). On photovoltaic electrolysis. *Alternate Energy Sources*, 3, 505–514.

Neill, D. *et al.* (1990). Hawaii Natural Energy Institute Wind–Hydrogen Program. In: *Proceedings of the 8th World Hydrogen Energy Conference on Hydrogen Energy Progress*, Hawaii, eds. Veziroglu, T.N. & Takahashi, P.K. pp. 71–77. New York: Pergamon Press.

NREL (National Renewable Energy Laboratory) (1992). *Proceedings of the 1992 DOE/NERL Hydrogen Program Review, Honolulu, Hawaii.* Report NREL/CP-450-4972. Golden, CO: National Renewable Energy Laboratory. (a) Mitsui, A., Biological hydrogen production, pp. 129–155; (b) Byliner, E.J., Photobiological production of hydrogen – solar energy conversion with cyanobacteria, pp. 157–167; (c) Weaver, P. *et al.*, Whole-cell and cell-free hydrogen for photobiological hydrogen, pp. 169–174.

Ogden, J.M. (1990). New prospects for solar hydrogen energy: implications of advances in thin-film solar cell technology. In: *Proceedings of Experts' Seminar, Energy Technologies for Reducing Emissions of Greenhouse Gases,* Paris, April, 1989. Paris: International Energy Agency (IEA) and Organization for Economic Cooperation and Development (OECD).

Ogden, J.M. (1993). Renewable hydrogen energy systems. *Solar Today*, Sept./Oct., 17–18.

Ogden, J.M. & Nitsch, J. (1993). Solar hydrogen. In: *Renewable Energy – Sources for Fuels and Electricity*, Ch. 22, eds. Johansson, T.B., Kelly, H., Reddy, A.K.N. & Williams, R.H., 925–1009. Washington, DC: Island Press.

Ogden, J.M. & Williams, R.H. (1989). *Solar Hydrogen – Moving Beyond Fossil Fuels.* Washington, DC: World Resources Institute.

Ohta, T. (1988). Photochemical and photo-electrochemical hydrogen production from water. *International Journal of Hydrogen Energy*, 13, 333–339.

Parmon, V.N. (1990). Photoproduction of hydrogen – an overview of modern trends. In: *Proceedings of the 8th World Hydrogen Energy Conference on Hydrogen Energy Progress*, Hawaii, eds. Veziroglu, T.N. & Takahashi, P.K. pp. 801–813. New York: Pergamon Press.

Plass, H.J. *et al.* (1990). Economics of hydrogen as a fuel for surface transportation. In: *Proceedings of the 8th World Hydrogen Energy Conference on Hydrogen Energy Progress*, Hawaii, eds. Veziroglu, T.N. & Takahashi, P.K. pp. 237–247. New York: Pergamon Press.

Rustamov, V.R., Abdullayev, K.M., Aliyev, F.G. & Kermov, V.K. (1998). Hydrogen formation from biomass using solar energy. *International Journal of Hydrogen Energy*, **23**, 649–652.

Shiga, H., Shinda, K. & Hagiwana, K. (1998). Large scale hydrogen production from biogas. *International Journal of Hydrogen Energy*, **23**, 631–640.

Veziroglu, T.N. & Takahashi, P.K. (eds.) (1990). *Proceedings of the 8th World Hydrogen Energy Conference on Hydrogen Energy Progress.* New York: Pergamon Press.

Wager, J.S. (1994). Renewables in the midwest. *Solar Today*, March/April, 16–18.

Winter, C.J. (1990). Hydrogen and solar energy–ultima ratio, avoiding a lost moment. In: *Proceedings of the 8th World Hydrogen Energy Conference on Hydrogen Energy Progress*, Hawaii. eds. Veziroglu, T.N. & Takahashi, P.K., pp. 3–47. New York: Pergamon Press.

Winter, C.-J. & Fuchs, M. (1991). Joint German–Saudi Hysolar Project. *International Journal of Hydrogen Energy*, **16**, 723–734.

Winter, C.-J. & Nitsch, J. (eds.) (1988). *Hydrogen as an Energy Carrier – Technologies, Systems and Economics*. Berlin: Springer-Verlag.

Wu, C.T. (ed.) (1990). AC–DC economics and alternatives – 1987. *IEEE Transactions on Power Delivery, Panel Session Report*, 5, 1956–1976.

Further reading

Anon. (1990). Solar hydrogen – pipedream of low-polluting fuel for the future? (News Focus). *Journal of Air Waste Management Association*, Jan., 86–87.

Beghi, G. (ed.) (1981). *Hydrogen: Energy Vector of the Future* (Proceedings of a Conference, 24/25 March, 1981). Haus der Technik eV, Essen, London: Graham & Trotman.

Bockris, J. O'M. (1980). *Energy Options: Real Economics and the Solar Hydrogen System*. New York: Wiley.

Kartha, S. & Grimes, P. (1994). Fuel cells: energy conversion for the next century. *Physics Today*, Nov., 54–61

Ohta, T. (1979). *Solar–Hydrogen Energy Systems*. New York: Pergamon Press.

Skelton, L.W. (1984). *The Solar–Hydrogen Energy Economy*. New York: Van Nostrand.

Stainberg & Cheng, (1982). *International Journal of Hydrogen Energy*, 14, 797–820.

Takahashi, K. *et al.* (1990). The Hawaii Hydrogen Plan. In: *Proceedings of the 8th World Hydrogen Energy Conference on Hydrogen Energy Progress*, Hawaii, eds. Veziroglu, T.N. & Takahashi, P.K. pp. 135–144. New York: Pergamon Press.

Veziroglu, T.N. (1987). Dawning of the hydrogen age. *International Journal of Hydrogen Energy*, 12, 1–2.

III The prospects for technological change toward sustainability

10 Summary assessment of the technologies

The prospective technologies for long-term supply of the world economy with energy on a sustainable basis are varied in kind, with widely differing promises and problems. There are, nonetheless, characteristics and challenges common to many of them. Uncertainty, as to the outcomes of the processes of research, development, and commercialization (RD&C) and their timeliness, is unquestionably a recurring concern with all the prospects that have not reached the stage of commercialization. The following chapter will review the RD&C processes, with their inherent promises and unpredictabilities along the road of technological innovation.

In the absence of public policies that promote renewable energy, we have assumed that market competitiveness will be the principle barrier to adoption for the candidate sustainable energy sources we have assessed here. In order to be competitive, these new technologies must have costs at or below those of the market. The unit costs of energy delivered must, generally, be compared with unit costs on world energy markets. For a new technology to be successful in mass displacement of conventional fuels or electricity, the competition must be assumed to take place with all of the conventional sources for which it is a substitute. If biomass alcohols, for example, are to compete and displace petroleum-based automotive fuels, then the unit costs of the alcohols must be competitive with those of oil-based fuels worldwide because of the universal transportability of liquid fuels. Much the same should hold for other sustainable energy sources whenever the product can be transported (e.g. gaseous fuels by pipeline) or transmitted (e.g. electric transmission line). Only in regional or niche markets would this general rule not hold up for meaningful price comparisons. Then, also, there are cross-elasticities between fuels (or electricity) whenever substitution is possible between sources, which will bring market pressures for price uniformity across countries and regions.

When we say that costs of these new fuels must be competitive with world fuel prices, we should not ignore the sources of uncertainty, in view of the history of world markets (Blair, 1978; Lee *et al.*, 1990). Whereas the world oil market is generally expected to be stable for the foreseeable future, the experience of the past – from the earliest days of oil – would tell us that the unexpected happens. Of course, if this were to be a *rise* in conventional fuel prices, it would make our alternatives more competitive, independent of their status in development. However, the past history of prospective alternative technologies shows that they can be marginalized through competition from *falling* fuel prices before they can penetrate their markets. It is this prospect that is the source of major anxiety for private investment in R&D and which suggests that only governments can take on the risk of such investments. It also means that the only alternative for the public policy in planning for such technological innovation is the creation of scenarios for market penetration that account for market-price futures (ETSU, 1994; WEC, 1995). Such a study is beyond the scope of the present work on the technologies themselves.

We have found these candidate technologies in differing stages of development. Solar technologies, for example, whether thermal or photovoltaic, continue to have the reduction of costs as the major challenge, although the solutions to this problem will be very different for the two branches of solar technology. Where those answers come from within the RD&C processes will also be very different, in view of the differing existing levels of technological sophistication. In the case of solar-heat collection, cost reductions will probably only arise from further straightforward development and design efforts, and discoveries from new fundamental research are unlikely to play a major role. By contrast, the development approach to PV cost reductions has not produced the results forecast by the industry and does not appear likely to do so in the foreseeable future. A return to fundamental research on semiconductors, optics, and thin films is required in this author's considered opinion. A prime example is in the area of the "copper-ternary" semiconductors, such as copper–indium diselenide ($CuInSe_2$), where issues such as chemical reactions in thin-film production and the dependence of semiconductor behavior on material composition are thought to be "rich in research opportunities" (Kazmerski, 1997).

Other renewable technologies, such as wind power and passive-solar collection, are already competitive in certain locations and situations. Further technological improvements, manufacturing innovations, and marketing efforts can be expected to enhance their penetration into their markets. Further penetration, however, will have limits for any of these new sources having intermittency, as with solar and wind generation. This inherent property will cause them to fall short of widespread displacement of conventional fossil fuels unless affordable means of energy storage are developed. The lack of economically feasible storage technologies to fill in for the periodic and random interruptions from these sources has already limited their adoption into integrated and stand-alone applications and can be expected to limit *accessibility* to them in the future.

The lack of economically feasible means of energy storage reflects a state of technological development that parallels that of the new source technologies themselves. Developments in advanced batteries, heat storage, and hydrogen storage have been anything but dramatic, even under the earlier impetus of the energy-crisis programs of the 1970s. Efforts to develop new batteries have produced only incremental improvements in life cycles and reduction of costs for stationary uses. Wind generation is the only intermittent technology where storage may be economically beneficial at the margin, given the incremental reductions that are forecast for the costs of advanced batteries.

Heat storage offers little but design refinements as improvements as long as it is restricted to using ordinary (sensible-heat) media. Taking advantage of the underexploited properties of phase-change materials seems to be a possible way out of this deadend. The three-step process for electricity storage in the form of hydrogen is, and can be expected to remain, prohibitively expensive as a means of storage. With any attempt to use these technologies, it must be recognized that the delivery of secondary energy from storage *requires additional investment, the fixed costs of which must be added to those of the primary-energy source.* Additional losses of energy are also incurred in conversion into and out of storage, thus making the unit cost of the energy delivered even higher.

The EVs, whether powered by batteries or by fuel cells, seem stymied by weight parameters that limit their range and speed. Developments in battery technology have progressed very little with regard to raising weight-specific power or energy storage of these vehicles. Ultracapacitors can, at added cost, enhance EV acceleration capabilities, but the range limitations of the basic battery packs will not be significantly improved. It is possible that new results will come out of the stronger emphasis placed by the recent Advanced Battery Consortium, but these too may be only incremental gains over time unless significantly different approaches can be brought in from basic research. The EVs using fuel cells (FCV) also have weight problems from the storage of hydrogen. Innovative alternatives, such as metal hydrides, are possible but further development efforts are needed. Even if weight problems are overcome for either of these electric vehicle possibilities, however, the present unfavorable cost comparisons with the conventional ICE automobile is another barrier to widespread adoption. (It should be kept in mind here that we are referring to *sustainably derived* hydrogen as a long-term goal and, therefore, are not including shorter-term proposals for hydrogen derived from natural gas in a chemical-reforming process on board the vehicle.)

Biomass energy resources have taken on a new significance for renewable-energy fuels with the emergence of new processing methods to produce alcohol fuels. These new processes permit the use of cellulosic feedstocks, such as wood and grasses, instead of only starches or sugars. In addition, thermochemical processes, previously used on coal feedstocks, can synthesize gaseous and liquid fuels. This has opened up a new, potentially vast, renewable energy resource base in the form of sustainable energy plantations of trees or crops. Further development and demonstration is required to assess the economic and sustainable feasibilities of the energy agriculture, including land productivity, but these efforts appear quite feasible to carry out, given adequate support.

However, in view of the massive scale of agriculture and silvaculture that would be required for biofuels, further research should be conducted on the impacts of monoculture environments, paralleling the development of feedstock production techniques. This would be prudent in order to avoid unanticipated outcomes (Tenner, 1996) to the ecology and biodiversity of surrounding environs. It will be important not to carry the mindset of contemporary agriculture into this new gargantuan undertaking (Shiva, 1993). A similar note of caution should be struck regarding biofuel processing in view of the massive scale of operations that would be required if these fuels were to replace significant amounts of fossil fuels. Here, the possible impacts on the environment, industrial health and safety, and social fabric should be studied carefully before embarking on such a mammoth industrial program.

Hydrogen as the fuel vector in a hydrogen-energy economy appears to be a somewhat farfetched idea for the foreseeable future, if for no other reason than its costs. This appears to be the case whether the primary energy sources are conventional or renewable. Certainly, solar–hydrogen has a doubly high cost barrier since it adds significant costs to the sustained high costs of solar electricity. The situation would be only somewhat better using wind as the primary energy source but would still be far from a competitively priced delivered energy.

Existing non-fossil-fuel technologies, like hydropower or geothermal power, offer limited possibilities for expanded exploitation, at least in most of the industrialized world. Large hydro projects in the underdeveloped countries, while technically feasible in select regions, are certain to encounter obstacles in financing and objections to their environmental impacts. Small hydro projects, however, have prospects for steady incremental gains in aggregate capacities in many different regions of the world, but the technically accessible resource has already reached its limits in several regions. Geothermal generation sites are confined to only a few regions of the world and can only have significant impacts on a local scale in those places.

It should be recognized that regional confinement is a characteristic common to most of the renewable energy resources. Hydropower and geothermal are examples of sources that are *both* regionally and site selective. Wind generation is somewhat in the same category. It is presently competitive only at "good wind" sites, which themselves are found only in certain geographical regions and then only where land usage will allow. Solar energy, whether through heat collection or PV generation, can only be expected to be competitive within the foreseeable future in the sunbelt regions of the world. The only technological prospect for overcoming such regional restrictions for mass uses of these energy resources is the use of recently enhanced capabilities of high-voltage electrical transmission to get renewable electricity to urban and industrial loads. Here too, however, concerns about impacts will have to be considered, as they have for large hydropower projects.

Biomass resources, unlike the other renewables, can serve to supply demand without inherent regional restrictions, provided liquid and gaseous fuels can be derived from them on a mass production basis. Once such fuels are developed to a cost-competitive status, they can be shipped or piped the way the fossil fuels are today and have a major impact on energy markets. The only regional consideration then becomes the cost of transportation, as with conventional fuels today. As such, they offer major prospects for world energy supplies, from wherever the land resources are accessible on a social and environmentally acceptable basis. In this sense, these regions (e.g. North America and Australia) can become the source basins for the flow of energy supplies in much the same way that oil and natural-gas basins are today.

Other renewable technologies that have captured the interests of researchers or inventors might be considered but have not been emphasized in our assessments here. These include ocean sources and energy derived from solid and liquid wastes, neither of which appear to have prospects for making significant impacts on widespread energy markets. While special applications and niche market applications can have localized importance, our objective here has been to assess prospects that have potentials for mass use, with realistic possibilities of making major penetrations into energy markets. Some of the prospects we have emphasized, such as solar–PV cells, have made inroads into niche markets but cannot be said to be consigned there permanently.

The promise of every one of these new technologies lies in the processes of R&D, which must precede their commercialization in the chain of technological change. We have reviewed the current status of each, regarding its technical workability and its costs in producing the commodity of energy. The outcomes of the R&D process will determine whether the workability and costs reach a competitive status. The progress of R&D, in turn, depends on its support from both public and private sectors. These factors, together with promotional policies in commercialization efforts, will determine which of the candidates makes it to the market and when. Chapter 11 reviews these aspects of technological change as they will apply to the sustainable energy sources.

The support for R&D in sustainable energy technologies can only come out of a public consensus in the industrial nations that alternatives to fossil-fuel energy must be found. This, in all likelihood, must be a political consensus, leading to government-sponsored research, since only governments can be expected to support long-term research with uncertain outcomes. The driving motivation for such public support should logically come from the recognition of the long-term risks of continued dependence on fossil fuels and the widespread benefits of a future with sustainable technologies to supply society's needs.

References

Blair, J. (1978). *The Control of Oil*. New York: Random House.

ETSU (Energy Technology Support Unit) (1994). *An Assessment of Renewable Energy for the UK*. Harwell, UK: UK Renewable Energy Programme, Department of Trade & Industry.

Kazmerski L.L. (1997). Photovoltaics: a review of cell and module technologies. *Renewable and Sustainable Energy Reviews*, 1, 71-170

Lee, T.H. Ball, B.C. & Tabors, R.D. (1990). *Changes in the Energy Situation: 1945–1989*, Ch. 2: Energy aftermath. Boston, MA: Harvard Business School Press.

Shiva, V. (1993). *Monocultures of the Mind. Perspectives on Biodiversity and Biotechnology.* Penang: Third World Network.

Tenner, E. (1996). *Why Things Bite Back – Technology and the Revenge of Unintended Consequences*, Ch.7, Acclimatizing pests: vegetables. New York: Alfred A. Knopf.

WEC (World Energy Council) (1995). *Global Energy Perspectives to 2050 and Beyond.* Vienna: the World Energy Council and the International Institute for Applied Systems Analysis.

11 Research and development

11.1 Introduction

In the earlier chapters, new prospective sustainable energy technologies have been assessed and the movement of some, such as wind generation, into the energy markets have been described. Others sources, such as biomass, appear to be promising but still need to move through the process of development to become marketable. Still others, such as solar PVs and batteries, are in serious need of basic research to advance to a stage of full technical feasibility and readiness for major energy markets. Finally, some technologies, such as those in solar thermal and thermal storage, are in need of further development efforts to get them to economic competitiveness.

For those technologies that seem to be presently limited, is progress likely within the foreseeable future? Can advances be made through R&D? If it can, how will this take place and within what time scale will it come to pass? What sort of financial support will be required; if it is provided, is there some reasonable prospect for viable alternatives in the next century? If funding support is provided for some particular technologies, can we expect that breakthroughs inevitably will result for those so supported, and if they do, can we assume that they will happen according to the neatly planned progression of development programs? The answers to these questions would seem to underlie any public policy making or private sector decision making that pertains to the prospects for new sources of energy in the future.

If we go a step further and assume that R&D efforts must be supported for technological change to take place, then we may ask: "What is the nature of this R&D establishment which will carry out these efforts throughout the world?" What are the specific characteristics of this establishment for *energy* R&D, as distinct from other areas of R&D? What can we reasonably expect from this establishment in the way of results and with what degree of timeliness, or even, with what degree of certainty of usable results at all for any particular technology? Finally, if outcomes from sponsored R&D are uncertain, is public funding necessary in the long term or can industry be assumed to invest in sufficient amounts to assure that there will be some technologies to replace fossil fuels when it does become necessary?

This chapter attempts to supply the energy-policy or decision maker with the insight into research and development to approach the answers to questions such as these, starting with the linear processes or stages of R&D, in terms of the programs followed and then in terms of the subtleties of the human interactions in the institutions where R&D is done. This is followed by a short survey of the scope of the energy R&D establishment, both publicly and privately supported, in the USA and other (mostly industrialized) countries where such efforts are found. Many of these aspects are discussed in editorial and comment columns in journals such as *Science*, and *Physics Today* and in newspapers such as the *New York Times*. Following this is a brief section mentioning the inner workings of such establishments, as seen by behavioral scientists and management specialists. The reader is referred to Appendix C which reviews the literature on observations of social behavior in the working situations of science and technology. These observations of group behavior show the effects of the social networks in scientific/technical research on R&D work and its outcomes. Finally, we come to the thorny questions of public priorities and the politics of government funding of research and development, with all of the uncertainties and frustrations of recent years.

11.2　The processes of R&D

In our assessments of the prospective new energy technologies, we have found them at various stages of development and with differing prospects for future adoption. For any of these case histories, R&D processes are involved both in finding paths to progress or in identifying the obstacles remaining to technological success. It should be useful to examine the workings of R&D in order to appreciate the inherent uncertainties and limitations of the process. These uncertainties and limitations apply not just to the ultimate outcomes of R&D efforts, but also to the timescale over which these outcomes might be achieved. Prime examples of the timescale factor are photovoltaics and their extended uses for solar–hydrogen (see Section 1.6 and Chapter 9). Our purposes here are, therefore, to look into the processes of research and development in general, and energy R&D in particular, including the institutional as well as scientific, technical, and financial aspects of how they are carried out.

Technological innovations have been recognized to result from an evolution through various stages of research and development: (i) new knowledge, (ii) development of technical capability, (iii) prototype or pilot operation, (iv) commercial introduction and (v) widespread adoption (Ray & Uhlmann, 1979; Tornatzky & Fleischer, 1990; Braun, 1992). Of these stages, the first two comprise the steps of R&D and the third has sometimes been called "demonstration". The entire process of bringing a new technology to marketability is, therefore, sometimes labeled RD&D (research, development, and demonstration). These stages are recognizable for the new energy technologies we have assessed here, although there is frequently a blurring of the distinctions between the steps as they are described in the press and even in the technical literature.

The first stage is where the fundamental scientific understanding of a possible working technical capability or where an entirely different concept of technical capacity may be recognized – this is the *research* phase. Many of the prospects we have reviewed appear to have passed this research phase in that the basic principles of operation are well established. Solar heating, for example, is based on physical principles of (solar) radiation, known for over a century, and heat transfer, known for over two centuries. Therefore, the research phase for the basic scientific principles of solar-heating technology would seem to be long past. The only remaining research outcomes possible would appear to be those in the area of materials for enhanced solar absorption or emittance, which have promise only for incremental gains in conversion efficiency or new materials for further reductions in costs.

The field of PV cells, however, on the other hand, appears still to be at a stage where research on basic semiconductor and optical properties of new compounds could result in major gains in the technology of solar cells. The fundamental science for PV conversion is in semiconductors, which is a young area of physics compared with those for solar-heat conversion. Semiconductors have provided dramatic technological innovations for other applications in the few decades since their discovery. Developments for PV cells, however, seem to have slowed before achieving electricity production that is competitive with mass conventional generation, despite the earlier optimistic projections of its backers. The gains reported for technical performance and cost reductions to date (Section 1.6) have been modest steps toward reaching this goal. What appears to be needed is a fundamental discovery or a more radical new approach (Ehrenreich, 1995), either of which is most likely to come out of basic research into the fundamentals of semiconductors and optics.

Biomass fuels provide an example of just how research can lay the foundation for later advances in development. The possibility of using cellulosic feedstocks for fermentation of fuel alcohol (Schell *et al.*, 1992) came out of earlier biochemical research on hydrolysis, saccharification, and fermentation of biomasses. Without this prior scientific inquiry into organic chemical reactions, development efforts for alcohol fuels would have been confined to starchy and sugar feedstocks, with all their inherent limitations (see Chapter 2). Prior research in this case had opened up the possibility of an entirely new set of technological options.

Nuclear fusion provides an example of a potential technology for which the basic scientific concept has *not* yet been proven for use and, therefore, cannot be considered as out of the research phase. Its predecessor technology, nuclear fission, moved out of its basic research stage when it was shown that fission chain reactions could be controlled. The basic science principles of fission reactors were in nuclear physics and thermal kinetics. For fusion reactors, the basics are nuclear and plasma physics. (A brief review of some of the technical barriers to achieving a fusion reactor is given in Chapter 8.) The working premise for a fusion reactor is that a net gain of energy can be attained from the energy output of the fusion reactions even after accounting for the energy input required to initiate them. The proof of this has yet to be made (Cassedy & Grossman, 1990). Critics of the fusion program have long contended that more emphasis should be put on questions of basic physics and less on program steps focused on results of the functioning of the reactor (Gilman, 1990; Maglich, 1991).

In all of these illustrations, there might be the impression that research is simply the first stage in a set of program steps that a new technology must be taken through. This is illustrative, as a matter of fact, of the confusion at large over the meaning of *research*. A very useful distinction in this regard between two very different conceptions of research has been made by Fox (1994) and others (Teich, *et al.*, 1994–5), who contrast "curiosity-driven" research with "strategic" research. The first is a more apt description of what has customarily been called *basic* research, whereas the second has been called *applied* research. Basic research has also been characterized as "investigator initiated". The underlying meaning of either this term or curiosity-driven is that the path of inquiry is not set at the outset nor is it imposed at any point along the way. It has been said that scientific research, particularly that conducted in the universities, is "removed from the application of knowledge . . . concerned instead with the extension of . . . knowledge for its own sake" (Mulkay, 1975).

Lest this is interpreted to mean an unbridled license to do just anything, it should be appreciated that research conducted by an individual or team receives its recognition through scientific or technical disciplines (Latour & Woolgar, 1979). The career of research scientists is seen to "center on the exchange of information for recognition" and the "ritual of publication" (Mulkay, 1975). The various disciplines provide critical peer review at publication on the validity and importance of research results. This provides guidance, which is at least implicit if not explicit, on the choice of the paths to be pursued by researchers in virtually every scientific discipline.

These processes of critical review are brought to bear through the institutions of the professional scientific and technical societies, with their journal publications and symposia. These professional societies have been likened to medieval craft guilds in their functions to assure the integrity of the work of the practitioners (Tornatzky & Fleischer, 1990). They are part of a broader "corporate structure" of science (Hagstrom, 1965; Brooks, 1968; Blissett, 1972) that provides a disciplined structure under which researchers must conduct their work if they are to get recognition for it. From such recognition follows funding for continued inquiry and professional advancement, thus feeding back to the drive for career ambitions within the universities and laboratories where professional careers are followed. Critics of the system, however, caution that the quality of the system depends on the sense of responsibility for society's needs in setting research priorities (Sarewitz, 1996).

Basic research has long been recognized as a public good in Western countries, with a tacit social contract understood between the scientific community and society at large (Brooks, 1993). In this understanding, it has been implicitly agreed that the scientific community has the freedom to define research objectives, allocate research resources, and, using peer review, select those who will carry out the research projects. It has been generally accepted that the scientists are seeking the fundamental principles and true essence of how things work in nature (Tornatzky &

Fleischer, 1990). Society has understood, again implicitly, that the benefits from this unfettered activity will be more than its costs. While this idealization of the arrangement has not always been realised in the political process, especially in recent years in the USA, it nonetheless has been an underlying understanding in modern Western societies. It was estimated that the total annual expenditure across ten US federal agencies for fundamental research in the early 1990s, in all areas of science and engineering, was $11B to $15B (DOE, 1991), with the total for the industrial world being approximately $40B (OECD, 1994).

In the biomedical field, the unfettered view of research has been recognized as an explicit policy by the US National Institutes of Health (NIH) (NAS, 1992). The NIH Office of Research Grants has emphasized the "integrity and independence of the research worker and his freedom from control, direction, regimentation and outside interference". Research proposals have been awarded on the recommendations of decentralized peer reviews in order to minimize outside political interference. The research results have also been subject to peer review for their publication in the various specialty journals of biology and medicine, following the practices of research fields in all branches of science. However, even though this unencumbered mode of allocation of public funds for basic research is generally accepted, it has been challenged by some observers of science (Sarewitz, 1996).

The ideal, nonetheless, is that the path to be pursued, in this basic research setting, is decided by the researcher; he or she is then free to follow it wherever it might lead and turn up whatever results that might be forthcoming. Such a setting has been proven to be one that has produced unexpected major advances in science, such as in the area of nuclear energy (Bush, 1945; Hagstrom, 1963). By contrast, the programmatic constraints of strategically directed research, or of the development process, can most often be expected to result in incremental advancements only, such as we have seen here with photovoltaics and batteries. Since these increments of progress can be forecast to occur with higher certainty than the major breakthroughs of research, directed research has come to be regarded as much less risky than basic research in appealing for corporate or government support.

The research underpinning technological advances has often taken place decades earlier and was not necessarily planned with that object in view (NAS, 1992; Ehrenreich, 1993). However, notions of basic research being unconnected with technological applications are themselves over simplified. Potential applications are usually foreseeable, at least in general terms, by researchers as the scientific results are emerging (Brooks, 1993), but the time scale for such development will usually be uncertain at that stage. Nevertheless, true basic research is more tied to scientific disciplines and not so much to application objectives; consequently, there is no apparent way to couple it directly into a program of strategic R&D, in order to achieve more timely results.

Useful results, leading to new technologies, are often said to occur from basic research by "serendipity" (Anon., 1993a; Schmitt, 1994) and, therefore, are not predictable as to where or when they will occur. Generally, what can be done, and usually is where true basic research exists, is to provide a scope of disciplines as the underpinning for a program of directed R&D. This is evidently the intended mode of the DOE, with its programs in Basic Energy Sciences, and has been to greater or lesser degrees with other federal mission agencies such as the DOD (US Department of Defense) in the past. It certainly has been the practice of the great industrial research laboratories of the USA in the past, such as the IBM Watson Research Center and the (old) Bell Telephone Laboratories.

Applied or *strategic* research, by comparison, has its objectives preselected in order to further a particular mission or development program. The investigators in strategic research are directed to pursue any or all ideas that they can conceive to advance on those designated objectives. They are constrained not to pursue ideas simply out of scientific curiosity, as the basic researcher might. In the photovoltaics program, for example, the "research" that is conducted is overwhelmingly applied to the objectives of the overall program for more efficient, more durable, and less expensive PV cells, working for a major part on modifications to already discovered materials and to a lesser extent on new materials. Similarly, for nuclear-fusion research where the first objective is to achieve a fusion reactor with a net energy output as a goal, all experiments have been conducted to further this end with various modifications of already conceived reactor configurations. Useful results, including some serendipidous ones, are likely to come out of applied research and in a more timely fashion than basic research. The advances, however, will usually be only incremental, as we have already mentioned.

Applied research is also said to be "technology-driven" rather than curiosity-driven and must perforce have an entirely different approach than to that of basic research (Tornatzky & Fleischer, 1990). The selection of problems is dictated by current technical needs, not by the unfolding threads of inquiry. The organization of projects, and even entire laboratories, centers around the technological objectives rather than scientific disciplines. The majority of applied research is carried out in industry and non-university laboratories. Over $100B is expended on applied R&D of all sorts in the USA annually, with nearly three quarters of that in industry and the remainder in the national laboratories, government-funded laboratories and (a small fraction) in universities. Similar expenditures for directed research, amounting to over $200B, are made in the remainder of the industrial world (OECD, 1994).

Our purpose up to this point has been merely to provide an understanding of what is loosely called "research", as it exists in the course of R&D for a new technology. We see that this step underlying the R&D process is highly uncertain, both in what results will be forthcoming and, importantly, in the timeframe with which any results will come. In Appendix C, we will comment further on the functioning of the research institution. At this point in the enumeration of the processes of R&D, however, we move on to the *development* phase.

Most of the prospective sustainable-energy technologies are in the development or demonstration phases, requiring either development of technical capabilities or demonstration of prototype operations. The results of prototype projects must first prove technical feasibility. Economically competitive operations must then be demonstrated, often on larger-scale pilot operations. Development is still required on all solar-heating (DHW, active space heating, IPH, and solar-thermal power) technologies, biomass, and energy-storage technologies. The field of PV cells also has had a number of development-type projects in which incremental technical and economic gains have been made. For several of these technologies, such as solar-thermal power and wind generation, we have described prototype projects that are approaching the "commercial introduction" stage mentioned above. In these efforts, there are programmatic steps taken to move systematically toward goals focused on technical workability and economic viability. There would be little time or patience for curiosity-driven inquiries or diversions into underlying questions in any of these development programs.

A good example of the development phase is the USA program for central-receiver, solar-thermal conversion in the USA. These projects, such as Solar One and Two, may be defined clearly as development and prototype demonstration efforts, even though they are sometimes promoted as the step leading directly to commercialization (Anon., 1997a). Each prototype model incorporates specific designs, materials and operating conditions, which represent combinations of state of the art features that it is judged will move engineering know-how a step closer to an operable overall design. Technical innovations, such as trying different coolant fluids or heat-storage media, are attempted from one model to the next. Some consideration is even given to the ultimate economic design, such as lowering the unit-area costs of heliostat reflectors. Some technical variations may be tried in the course of the testing for a given model. However, no deviations from the pre-agreed, basic modes of operation would be expected during the course of any development program such as these.

Wind power is an example of a technology that has moved in large part through the development and prototype stages and is now entering the commercialization stage. As we have described in Chapter 3, wind-turbine technology has evolved to its present state of operability and cost competitiveness partly through (government-sponsored) development (DOE, 1997) and partly through wind machines commercially developed in both the US and in Europe (Gipe, 1995). The development of these machines under private sponsorship was motivated, in turn, by the success of the incentive-driven California wind farms and the Danish agricultural market. The use of wind power has illustrated how a new technology, based on well known scientific principles (mainly, aerodynamics) can be carried through a (difficult) development phase and move into commercialization.

Biomass, by comparison, is an example of the development process still very much in progress. Virtually all of the promising features of this technology have still to be integrated into workable operations and designs with good prospects for economic competitiveness. Operations development starts with the growing and harvesting of feedstocks, be they energy crops, grasses, or wood. As outlined in Chapter 2, the unit costs of the feedstocks must be held to sufficiently low levels, along with processing costs to produce fuels at competitive costs. In addition, the agricultural methods must be put on a sustainable basis. Finally, chemical processing must be designed for mass production and the prototype operations proven. All of this will require an extended development program, integrating the various processes into a working, economic whole. The DOE (1993) has projected their program extending well into the next century.

The RD&D process is only part of the overall path of *technological innovation*, which is the entire technical *and* social process leading to the widespread adoption in society of a new technology. Any of the new energy-source technologies that do reach the stage of commercialization will still face the final barriers to widespread adoption into the market for its energy (Rogers & Shoemaker, 1971; Girifalco, 1991). Even with a functioning and economic technology, barriers still exist to adoption. The classic example for renewables is SDHW, as mentioned in Section 1.2 (FSEC, 1979). During the 1970s subsidies made SDHW economic in sunbelt regions but widespread adoption did not follow close behind. Even with substantial financial incentives, market penetration was slow and hesitant, because of social, psychological, and institutional barriers. It is generally accepted following well-known sociological behavior patterns, that breaking social barriers requires building a network of confidence in a new technology or new way of doing things.

The process of diffusion of a new technolgy must be initiated by risk-taking "early adopters". Psychological barriers also will exist, such as the "trialability" in the use of a new and unknown technology. Also, the establishment of trustworthy guarantees, performance standard and repair services are essential, as sad experience in early SDHW systems illustrated (FSEC, 1979; SRRC, 1993–4). Finally, legal conflicts over the rights to solar access created institutional barriers to investments in solar collectors. Land-use covenants are expected to play a large role in the adoption of wind farms (Brower *et al.*, 1993). All of these non-technical and non-financial obstacles must be overcome if a new technology is to achieve its market potential.

11.2   The R&D establishment

The establishment that carries out energy-related research and development varies in the USA and other countries having such programs. In the USA, energy R&D laboratories are both government run and privately run, with government funding going to some private labs. The federal government labs doing energy R&D are certain of the national laboratories, for example the National Renewable Energy Laboratory (formerly the Solar Energy Research Institute), Sandia National Laboratories, Los Alamos National Laboratory, the Idaho National Engineering Laboratory, Lawrence Livermore National Laboratory, Argonne National Laboratory and Oak Ridge National Laboratory (all under DOE), and the

National Institute of Standards & Technology under the Department of Commerce. Several of these national labs also monitor energy R&D, related to their objectives, on federal research contracts and cooperative agreements with universities and non-government labs.

In Europe, R&D in all areas is carried out on an OECD basis, by national research agencies, and by private industry. The sponsorship of energy R&D is an integral part of the energy policies of the EC, as defined by the Maastricht Treaty, to provide energy security, energy efficiency, and develop renewable technologies (Lyons, 1994). The announced energy policy of the EC to meet environmental objectives includes short-term and long-term goals for development of renewable sources (Cross, 1993). These policies have been reinforced by a subsequent declaration of intent (EUFORES, 1998). Of the funding from within each EC country, somewhat more than half on average is funded by industry, a similar situation to that in the USA (Nicklas, 1997). Government funding comes from ministries of energy, environment, industrial development, and the like that have missions relating to industry. The average national R&D expenditure, on research of all types, for OECD countries is about 1.75% of gross domestic product (GDP), compared with about 2.75% for the USA. Out of these expenditures, energy R&D (of all types) accounts for only about 0.03% of GDP in the USA and several of the European countries. The Netherlands expends over 0.05%, but that rate appears to be dropping. France's energy R&D funding was 90% for nuclear projects and Spain's is mostly for fossil-fuel research. Finland expends about 25% on renewables, over twice the percentage of the USA for energy R&D. For energy R&D, each European country seems to concentrate on selected technology areas, such as wind, solar–thermal or electric vehicles (IEA, 1996b).

Renewable-energy projects have sprung up in most of the OECD countries in Europe and in Japan. By 1990, an aggregate of almost $250M in funding for renewables had been allocated by Germany, Italy, the UK, Greece, the Netherlands, and Denmark (Anon., 1991, 1992). Japan had allocated nearly $100M to renewables. The Japanese Sunshine Project is devoted to projects on solar, wind, biomass, geothermal and ocean-energy technologies. The UK allocated about $30M in its 1993 R&D budget to renewables, including biofuels, solar, wind, hydro, tidal, wave, and geothermal projects (IEA, 1996a). France budgeted about $17M for renewables R&D in 1992 (Cross, 1993). Overall, however, the aggregate of OECD governments R&D expenditures for renewables has fallen from an "energy-crisis" high of $1202M in 1982 to $487M in 1992, rising to $694M in 1993 (Nicklas, 1987). Projecting ahead, allocations have been made in Germany and the Netherlands for wind installations by the year 2000, $238M for 250MW capacity and $1050M for 1000MW, respectively (Evans, 1992). In Denmark and Italy, capital-cost or operating incentives for wind installations have been instituted, including waivers for VAT and carbon taxes for wind-generated electricity (Gipe, 1995).

Cooperative R&D efforts were initiated in the 1970s in the EC countries, looking forward to the technological integration and innovations needed with the impending economic integration. A salient example of these efforts is the Framework R&D Program, which has been funded at $8.4B for 5 years for work in communications, materials, biotechnology and energy (NAS, 1992). Another is the THERMIE Program, sponsoring projects in renewable energy sources and energy conservation. Also amongst the EC programs is EUREKA, involving 12 OECD member states, with 120 projects in all areas of R&D, including 34 on the environment and four in energy (Anon., 1991, 1992). More specifically relating to renewable energy sources, the JOULE (Joint Opportunity for Unconventional or Long-Term Energy Supply) Program sponsored research in thermochemical processes for biofuels expended almost $140M over the period 1989–92. It continued into further programs in solar, biomass, geothermal, hydro, and wind, expending upwards of $800M to 1994. Also, the ECLAIR (European Collaborative Linkage of Agriculture and Industry) funded over $90M to a biorefinery project over the period 1988–93 (Cross, 1993).

The IEA, operating under the auspices of the OECD, has over 360 implementing agreements with member countries for collaboration on energy projects, with over 85 of these relating to renewable energy sources (IEA, 1994). Six of these projects on renewables were with US facilities. The EC's Research Budget for 1992–94 included $1276M for all energy R&D, out of which $320M was for "non-nuclear energy", $274M was for nuclear fission safety, and $682M was for controlled nuclear fusion (*Science, Special Issue*, 1993).

US federal funding for energy R&D has come mainly, of course, from the DOE. All federal funding was subject to deep cuts in the Congressional sessions of 1995–96. Since either further cuts or no significant increases in DOE funding can be expected with the current composition of the US Congress (Anon., 1997b, 1998), our purpose here will be to give a view of allocations at their highest point prior to those cuts, thus reflecting priorities up to that point.

Total DOE R&D funding was approximately $7B annually in the early 1990s (AAAS, 1994), prior to the budget/tax debate in Congress. Out of this overall allocation, nearly $3B remained allocated to nuclear weapons issues, despite the end of the cold war. This had left about $4B for categories such as energy supply and conservation. The DOE Energy Supply Budget category (fiscal year 1995 $2.22B requested) included major subcategories of interest in this book, such as:

solar energy: $258.3M budgeted
geothermal: $ 36.2M budgeted
hydropower: $ 0.9M budgeted
hydrogen: $ 5.5M budgeted
energy storage: $ 46.9M budgeted.

Biofuels, wind, and ocean systems R&D were included in the "solar energy" category in addition to solar-building, solar-thermal systems, and photovoltaic projects. Some of this funding was for cost-sharing projects with industry to aid market penetration of PVs, thus reversing to a small degree the earlier policies of non-involvement in commercialization. Total funding for PVs (out of the overall solar funding) was about \$85M in 1995, with cost-sharing being a fraction of that (Williams & Bateman, 1995). An estimate made before budget cuts for the entire solar and renewable energy R&D, including some funding not shown under this category, showed \$397.5M budgeted for FY95. While an increased emphasis on renewable sources R&D had been advocated from elsewhere in the US government (GAO, 1993) and US professional/industry groups (IEEE, 1994), no increases in funding were anticipated, even in that period.

Separate from this funding of energy supply (sources) was a category of energy conservation, which had FY95 requested funding of \$502M. This was to be devoted to subcategories of building, industrial, transportation, and utility conservation technologies. Also separate from these budgets was that for nuclear fusion research, funded at about \$325M in FY94, which had already been cut as discussed at length in *Science* (Anon., 1993b–d, 1994a). In addition, there was a separate category for Basic Energy Research (BES, total FY95: \$731.4M), which funded a range of basic science defined by broad disciplinary categories such as chemical sciences, energy biosciences, engineering and geosciences, and material sciences. These four discipline areas had an aggregate FY95 budget of \$499M, some of which (if funded) could have produced outcomes relevant to the development of the sustainable, energy-source technologies of interest here. About one quarter of the total BES funding was to go directly to university-based research. All told, the annual federal expenditures for fundamental research in all agencies was about \$11B prior to the budget cuts of 1995–96 (DOE, 1991).

During the oil crisis period of the 1970s, several of the US states initiated energy policies to meet the perceived needs of their constituencies. Some established energy offices to administer incentive programs. Forty two of the fifty US states had incentive programs of one sort or another for renewables in 1997, including tax breaks and loan programs (Pye & Nadel, 1997). Twelve of the states were giving financial assistance specifically to the commercialization of photovoltaics (Williams & Bateman, 1995). Others have gone beyond office bureaucracies and established agencies to promote alternative energy technologies and energy conservation. The best known and longest lasting of these state agencies are the California Energy Commission (CEC) and the New York State Energy Research & Development Authority (NYSERDA). The CEC has been a pioneer among state agencies in the USA for sponsoring research, incentives, and influencing policies within a state. It was founded in 1975 in response to the acute air pollution in the Los Angeles air basin (van Vort & George, 1997). The state's requirement for a future "zero emission vehicle" has stimulated R&D in the battery and automotive industries across the US.

NYSERDA was founded in 1975 to "help secure the State's future energy supplies, while protecting environmental values and promoting economic growth" (NYSERDA, 1997). It has sponsored (or co-sponsored) energy-related RD&D programs ranging over topics such as industrial processing, energy conservation, demand-side management, municipal waste management, and nuclear waste management, as well as alternative energy sources. In the 1993–94 fiscal year, the NYSERDA R&D budget was $17M, out of which about $7.3M was allocated to energy resources, including over $2.5M for projects related to renewable energy. These renewable-energy projects included PV demonstrations, PV semiconductors, solar monitoring, novel wind generators, and biomass/ethanol processing. In the previous 5-year period, over $8M was expended on biomass projects, including wood pyrolysis, tree farming, cellulose/ethanol, and methane from energy crops. Many of these projects were conducted at universities on a basis comparable with federally sponsored R&D. With the budget/tax debate carrying into state capitals, however, it is likely that support such as this will suffer the same fate, or worse, than federal funding. For example, the NYSERDA renewable energy budgeting dropped from $2.5M to $1.8M in the 1997–98 budget.

Nineteen state energy agencies have been formed in addition to state-based public-benefit organizations devoted to innovation in energy technologies in the USA and these belong to the Association of State Energy Research and Technology Transfer Institutions (ASERTTI) (Pye & Nadel, 1997). ASERTTI was formed in the face of cutbacks in federal R&D funding and utility restructuring to promote research and information dissemination at the local level. The majority of their projects focus on energy efficiency and technology transfer results from federal and industrial projects. There are some development projects on renewable sources, such as those mentioned for NYSERDA, but the emphasis is on technology making local impacts (e.g. SDHW). There was over $65M in RD&D funding in these states in 1995–96, with matching funds over $100M from the private sector.

There are several non-government, not-for-profit labs, foundations, and industry groups doing energy R&D in the USA or sponsoring it. A partial list of those working on renewable energy technologies and related research includes the Electric Power Research Institute (EPRI), Institute of Gas Technology (IGT), Pacific Northwest Laboratory, Battelle Memorial Institute, Research Triangle Institute, Energy Research Corp., Southwest Technology Development Institute, Center for Semiconductor Research, Solid Waste Association of North America, the Solar Energy Industries Association (SEIA), and the Utility Wind Interest Group. Some of these are sponsored by specific industries, such as EPRI and IGT, which are funded by public utilities across the USA.

Energy R&D is also carried out by universities. In academic departments in the sciences and engineering, basic research of all types has been supported by the National Science Foundation (NSF) as a matter of national policy related not only to the nation's research effort but also to support graduate (doctoral) education (Teich *et al.*, 1994–5). A total of $2.28B was budgeted in FY95 for all "research & related activities" (discussed in *Physics Today* (Anon., 1993e, 1994b,c)), covering all areas of research and graduate programs including energy. (This NSF funding has actually increased somewhat since the budget battles began; the funding had increased to $2.55B for FY98 (Anon., 1997c).) The NSF research programs have been organized along broad disciplinary lines such as biological sciences, computer and information sciences, engineering, geoscience and the like, not along the lines of technological applications as development projects would be. Therefore, just as with the DOE Basic Energy Sciences budget, it is not apparent what the NSF allocations relevant to energy technologies might have been. However, it has been estimated that over $600M had been going annually to universities from the DOE for energy R&D, with most of that being for development projects not basic research. Not all of these are solar or renewable-energy projects, of course.

Several of the research-oriented universities in the USA operate research laboratories separate from their academic departments. In some cases, universities operate a national laboratory, for example the University of California (the state-wide system) operates Los Alamos and Livermore. Other universities, however, operate wholly owned laboratories that are not for profit. Within this group of university labs, R&D related to solar and renewable sources has been carried out at academic centers such as the Atmospheric Sciences Research Center (SUNY), the Florida Solar Energy Center (University of South Florida), the Georgia Tech Research Corp. (Georgia Institute of Technology), the Hawaii Natural Energy Institute (University of Hawaii), the Institute of Energy Conversion (University of Delaware), the Jet Propulsion Laboratory (California Technical), the Lawrence Berkeley Laboratory (University of California) and the Purdue Research Foundation (Purdue University).

Private industry in the USA spends more on R&D (of all sorts) than the federal government (AAAS, 1994; DOE, 1994). It was estimated for 1993, for example, that industry spent $81.3B (of corporate funds) compared with the government's $68B. If we add government R&D funding to industry, we find an additional $31B. This picture was little changed in 1997 (Anon., 1997d). Industrial R&D is concentrated in a relatively small number of companies, with 79% of such expenditure in the USA made in 100 firms (Tornatzky & Fleischer, 1990). While no good estimate is available on industry's energy R&D funding, we have seen that the government's funding in this category was about $4B or 6% of its total R&D prior to the funding cuts of 1995–96.

There has been a rather long list of private, for-profit companies in the USA working on renewable energy devices or systems or that have been involved in government-sponsored projects. The majority of this private, industrial effort is on the development area of R&D, devoted to solar, wind, hydrogen, and related technologies (NAS, 1992). Some of the sponsored energy projects evolved through traditional R&D contracts while others are carried out in the new form of Cooperative Research and Development Agreements (CRADAs). The CRADA format was created in the USA to capitalize on the R&D capabilities of the national laboratories and help to move new technologies into commercialization with private business. This was part of a broad US federal program to improve the nation's industrial competitiveness (Brooks, 1993). Still others in the USA have been able to operate, either with the benefit of government incentives or even without, to supply energy from renewable sources such as wind and hydro (Williams & Bateman, 1995). However, as a consequence of varying and uncertain government support, some of these, such as LUZ (solar) and Kenetec (wind), have gone bankrupt.

In Europe over 3700 companies have been listed in a directory of "renewable energy suppliers and services" (Cross, 1993). Out of this number, about 500 can be identified with the technologies we have reviewed in this book, such as solar collectors, ethanol, anaerobic digestors, thermal gasifiers, photovoltaic systems, and batteries. Most of these companies are supported with the aid, direct or indirect, of the various governments or the EC, as discussed above.

Electric utilities have been involved in the USA, often with (federal or state) government incentives, in prototype demonstration projects in solar-thermal, photovoltaic, wind, and hydroelectric generation, as we have seen in our reviews of these technologies. Some of these utilities include Central Maine Power Co., Connecticut Light & Power Co., Green Mountain Power Corp., Hawaiian Electric Industries Inc., Louisiana Power & Light Co., New England Power Service Co., Niagara Mohawk Power Corp., Northern States Power Co., Pacific Gas & Electric Co., Pasadena Water & Power Department, Washington Water Power Co., Municipal, and the Wisconsin Electric Power Co. In the case of wind generation, operations have continued beyond pilot demonstrations to the ongoing contractual delivery of electricity to utilities. The prime example of this is the collection of wind farms in the Altamont Pass of California, selling electric energy on a routine basis to PG&E. Other examples of wind projects include Puget Sound Power & Light and the Bonneville Power Administration (in the Pacific Northwest), Idaho Power (in the West), and Sacramento Municipal Utility District (in California). (SMUD is also involved with central and distributed PV generation, see Chapter 1.) There are indications that the onset of deregulation and competition will result in US utilities investing in proven technologies only. In the early 1990s, for example, some utilities scaled back their PVUSA demonstration projects (Williams & Bateman, 1995). The National Association of Regulatory Commissions has also expressed concern that levels of support for energy R&D are dropping under the pressures of competition (Anon., 1997e).

Various government programs have been established (and dis-established) in the USA to aid in the development and commercialization of alternative energy technologies. Many have been incentive programs directed at commercialization, for example CRADAs, which target the development phase for the involvement of private industry. Another government-industry activity for alternative energy technologies is the Committee on Renewable Energy Commerce & Trade (CORECT), which was established by Congress in 1984 to promote the use of US renewable energy products and services around the world. CORECT is a working group of 14 federal agencies, including the DOE, whose main mission is to facilitate technology transfer in the use of US-developed renewable technologies, principally in the Third World. The activities of CORECT not only serve the announced purposes but also give an added promise of early commercialization for the technologies under development and further support for the US export trade. Gipe (1995) has pointed out that Denmark created an export market for its new generation of wind machines, doing so after spurring a "market pull" within their own country for this new technology.

11.3   The management and conduct of R&D

The earlier part of this chapter has described the purposes, organization, and current support of the research establishment upon which the long-term hopes for new, widespread sustainable energy technologies rest. The sponsors of research and development have a stake in that sponsorship, whether public or private. In the public domain, the social contract carries the assumption that results useful to society will come out of the allocation of public funds. In the private sector, shareholders expect returns on their investments in R&D. Consequently, the question is raised as to what results can reasonably be expected from investments (private or public) in R&D and on what timescale and with what certainty can those results be expected to be forthcoming?

The answers to these questions are not simple, because the functioning of the R&D establishment is anything but simple. *Uncertainty* is the watchword throughout the process of technological innovation: uncertainty as to both the outcomes and the timescale for achieving those outcomes if they are realised. The policy maker or analyst can best make judgements by getting some insight into how the R&D process functions.

Appendix C is a review of the literature on the conduct of basic research, development projects, and the management of (industrial) R&D. It goes into the behavioral aspects of the groups and individuals taking part in these activities, attempting to illuminate where the uncertainties, previously alluded to, enter the process: the paths of inquiry taken, the uncertainties of outcomes, and the unpredictability of time of outcomes. There are open questions regarding the optimum size and management style for development projects. An appreciation of these subtleties is essential for anyone making decisions or policy recommendations for the support and planning for R&D.

11.4   The politics of R&D

As much as the successes or failures of the R&D process may be determined by the conduct of the work itself in the laboratory, the outcomes are still very much at the mercy of the forces of the political process external to the laboratory. The political process in the industrial democracies is embodied in the legislative process taken together, but not always in harmony with, the policies of the executive branch of government (when that is separate, as in the USA). The forces at play range from underlying ideology to particular special interests and can make for a highly uncertain environment for planning and conducting R&D programs with long-term objectives, as experienced, for example, in the USA (CCS, 1994).

Our earlier listing of the current R&D efforts and budget allocations for new energy sources gave an indication of the size of the establishment in the industrialized world, but it gave no indication of how this was achieved within the political process. The allocations reflect, of course, the priorities of national policies coming out of Congress and the White House in the USA and the parliaments/governments of the other industrial countries. Energy policies have universally suffered from wild changes and uncertainties since the 1970s. Wide swings in energy markets (mostly in oil) and major changes in ideology regarding matters of energy and environment have generated discontinuity and unexpected changes in energy budgets and planning during this period. Not only have there been shifts in allocational priorities within energy budgets but there have also been incessant pressures on government spending overall. In the USA these pressures result from the federal deficit and political rhetoric against taxes. Budget rhetoric seems more strident as we move toward the turn of the 20th century and at a period when we are looking for innovations to move us toward sustainability. Any assertions that funding basic research will accrue to the long-term benefit of the public would appear to be lost in this political climate.

The political debate on public support of research has also, for some time in the USA, turned on the views as to which approach – basic (unfettered) or strategic (directed) research – is most likely to produce results for acknowledged national missions. The dichotomy has been said to be between belief in "serendipity versus strategy" (Schmitt, 1994). The differences in judgment between the scientists lobbying for research funding and the politicians legislating that funding is determined, of course, by the very different perspectives and experiences of the two groups. The tenuous position of the scientific community has been summarized by Brooks (1971): "Although science cannot ask for a blank check, there is a part of it which must have autonomy to 'do its own thing' if it is to serve society . . . (however,) some accountability . . . is essential . . . the degree (of which) . . . will depend . . . on (how) . . . science maintains its own system of internal accountability, guaranteeing the excellence and integrity of results." This debate has continued in the USA since World War II (Smith, 1990) and continues into the period that has followed the Cold War (Anon., 1993a; Kelly, 1994). The trend, starting in the 1970s, has been more and more ". . . to get our knowledge fully applied . . . harnessing science . . ." (attributed to President Lyndon Johnson) (Remington. 1988). The legislative directions in funding have been discussed regularly in the journal *Science*: in recent years funding has moved more to favor *strategic* rather than *basic* research in the USA (Anon., 1994d–g) and in other countries as well (Anon., 1994h), although it is still the subject of debate (Anon., 1994i–k).

Even though scientific research may have been viewed abstractly as a generic "public good" by respected public figures (Bush, 1945; Brooks, 1993), others have observed that support of science in the USA, Great

Britain and France appears historically to have come about only with the sense of urgency that accompanies widely accepted national missions (Dasgupta & Stoneman, 1987). These missions, such as defense, space, and health, have been based on political priorities . Other industrial countries, such as Germany and Sweden, have tended to support R&D on less categorical bases – rather more generally in areas likely to aid domestic industries. Research as underpinning for new energy technologies in the USA and other industrial countries has been limited mostly to nuclear (fission and fusion) and fossil-fuel technologies. These allocations have been in support of broad national missions for defense, electric power, and transportation but also reflect the heavy political influence of those industries. More recently, with regard to energy R&D, European nations have agreed on rationales for govenment intervention in energy markets, such as providing energy (fuel-supply) security, environmental protection, and provision for long-term energy alternatives through R&D (Lyons, 1994; IEA, 1996c).

Japan is a prominent example of cooperative R&D with industry by a central government to further industrial products (NAS, 1992). Impressive results, such as the 4th Generation Computer Systems, have been achieved in programs sponsored by the Ministry of International Trade and Industry (MITI). The R&D is carried out both in the industrial labs and in Japanese national laboratories, which is a model of what can be accomplished in this mode of support, albeit in a different area of technology.

All of this, however, raises a policy question: to what extent should the government be involved in the processes leading to adoption of a new technology into the market? In the USA there has been an extended policy debate, beginning in the 1980s, on the appropriateness of government support extending beyond the research phase toward commercialization (NAS, 1992). The position of the Reagan and Bush administrations was that development and commercialization should be left to private initiative, as governed by the market. New energy technologies, in particular, received little support to accelerate adoption in the market during that era. The SDHW demonstrations of the 1970s were dropped in the early 1980s because of this political stance. Both wind and solar PV have moved into their demonstration phases only with the aid of financial incentives from state governments as well as the federal agencies in the USA and not merely as federal-sponsored RD&D projects.

Prior to 1995, trends in the USA had shown increases in the federal allocations for solar and renewables and decreases for nuclear (fission) and most fossil-fuel categories (one exception is natural gas) (AAAS, 1994). Nuclear-fusion research had received only small increases and any initiation of new major experiments in US laboratories seemed to be questionable even then. These shifts reflected political trends resulting from an increased public awareness of energy/environment issues, although this too may not continue (Anon., 1994i,j, 1995). In the USEPA, for example, these trends have even contributed to an increased emphasis on long-term research on environment-related areas of science (Anon., 1994k).

Starting in the early 1990s, even before the change in US administrations, a new debate on the role of the later phases of the R&D processes has taken place (Dasgupta & Stoneman, 1987; de la Mothe & Ducharme, 1990; Florida & Kenney, 1990; Rosenberg *et al.*, 1992). This debate was driven by the concept of new technology as a source of economic growth overall and as a means to enhance national competitiveness. This situation already operated in the industrial countries in Europe and in Japan. The Clinton administration extended these themes to the level of industrial policies supporting not only R&D for the stimulation of US industries but also R&D that would anticipate and stimulate market demands; a topic frequently commented upon in the journal *Science* (Anon., 1993b–d, 1994a). The thrust of these policies was to help to create new products and increase industrial productivity. It was clear, however, that such industrial policies would run against the free-market philosophies of the Republican delegation in Congress (Anon., 1994i,j, 1995).

In the broader sense, however, questions of international competitiveness are not central to policies for energy R&D, since energy resources are in the nature of an infrastructure for the entire economy of an industrialized country rather than manufactured products that can compete on the international market. Presently, the USA, Europe, and Japan are net importers of energy to meet the infrastructure needs underpinning their industrial economies. This might lead one to conclude that alternative sources of energy might be sought. This, however, has not been the case to any pronounced degree in the USA and no high priorities have been attached to it in the political arena there, the advocacy of the Vice President (Gore, 1993; Anon., 1993f) not withstanding. There has been more awareness of these vulnerabilities in Europe and Japan, dating back to the 1970's oil crises. Responses to this concern have shifted in most of the rest of the OECD away from nuclear alternatives toward renewables, with added impetus from environmental and climate concerns.

Whereas there are prospects for export of some renewable-energy products, such as solar–PV systems, the major national need for the USA and other industrial countries is for massive sources of energy other than fossil fuels. It would appear that the central mission of energy R&D in these countries must be conceived as distinct from the "competitiveness" mission (NAS, 1992; Branscomb, 1993; Teich, *et al.*, 1994–5). The national mission for energy R&D, if one is to be formed, would have to be a focused one, in the nature of national space or health programs. The mission would be *to provide alternative energy sources ultimately capable of supplying energy on a mass scale sufficient to displace most use of fossil fuels in the national economy.*

It should be made clear here that we are talking about programs to develop new *sources* of energy to displace fossil fuels on a large scale in the long term, not merely trying to slow or reduce consumption of these fuels. We are referring to extensive R&D programs for technological innovation of these new source technologies, not merely relying on short-term programs using policy instruments such as tax penalties (e.g. carbon taxes) or incentives (e.g. credits or deductions for energy efficiency) to reduce the consumption of fossil fuels (Roodman, 1997). Such a policy would have to be a bold new departure, with a full commitment to replacing fossil fuels, albeit in the long run. Such a policy would also be a prudent response to the warnings that world demand for ever increasing supplies of energy will inevitably result in more greenhouse emissions, even with strenuous efforts at energy efficiency and conservation.

Currently, it would appear that the only prospects in the USA for major increases in R&D for renewable energy sources would have to come out of a newly conceived national mission for an all out attack on environmental degradation and climate change (Hollander, 1992; Rosen & Glasser, 1992). This could come about only with a public perception of crisis, such as was engendered by the oil embargoes of the 1970s (Freeman, 1974; Gever *et al.*, 1986). That sense of crisis, of course, was short lived and did not result in a sustained political will to work for long-range solutions. It also did not focus on the environmental impacts of energy sources alternative to oil, as clearly indicated by the emphasis then on projects for fuels derived from fossil resources such as coal and shale oil (Stobough & Yergin, 1983; Yanarella & Green, 1987). While attitudes in the US have shifted away from fossil-derived fuels as the source of alternatives, as a result of increased environmental awareness, no national consensus of mission for the climate and the environment has evolved in the country's politics.

Despite this, however, advocates for renewable energy have posed scenarios of bold programs of development (Nicklas, 1997). With sufficient investment, they project growth in the contribution of renewables to primary energy demands in the USA rising from the levels of under 9% at the end of the 1990s, to over 25% by the year 2010, and over 40% by 2020. They pose scenario cases of environmental impacts and policy determinations for climate stabilization as the driving motivations. The Renewable Energy Advisory Group of the UK recommended in 1992 that up to 25% of electricity be supplied from renewable sources by the year 2025. A German study suggested a possible 13% contribution by solar energy alone by 2005, rising to levels over 45% by 2025. This study projected the required solar investments to be at $7.5B per year through 2005 and over $30B annually for the two decades that follow. Such projections are well intentioned but unrealistic given the current outlook of governments and industry in the industrial world.

Even if sizable sponsored R&D projects were initiated, however, there would still be the controversial question of commercialization – whether such projects should receive government support. History records many failures of new technologies to penetrate their intended markets for reasons beside their technical readiness. As mentioned previously, transitional subsidies or tax incentives have made the difference between success (wind farms) and failure (solar–thermal fields) in the recent past for new energy technologies (Anon., 1997a). Such programs, however, are not in favor during the present era of government budget cutting and free-market philosophy in the USA. By 1985, the USA had phased out most of its "energy-crisis" era commercialization programs, which were started in the mid-1970s. This policy shift was an explicit expression of the private enterprise philosophy of the Reagan–Bush presidencies, proclaiming that the government should support development of new technologies only through the research phase, leaving development and commercialization to private industry.

Several European countries in recent years, however, adopted programs of price subsidies to aid in the market transition for the new technologies, such as wind, that were nearly competitive. In the USA, initial cuts in national price supports for commercialization of renewable technologies were offset by reinstatement of small production credits and construction credits in the *Energy Bill* of 1992, which were allowed to lapse again in 1998. The consequent irregular and uncertain nature of support contributed to the financial failures of projects such as SEGS (see Chapter 1). The most consistent support in the US has been from the State of California, contributing to the temporary success of the SEGS projects and the continued success of the California wind farms (see Chapter 3). These winds farms, while significant demonstrations, are modest in terms of energy production and represent only a regional niche in terms of market penetration.

Another force stimulating these new technologies, however, may be the evolving restructuring of the utility industry, which has grown in part out of environmental concerns (Weinberg, 1994). These trends seem to be toward smaller, cleaner, and more dispersed electric generation plants, with less "lumpy" concentration of investment. The modular nature of renewables, such as wind or solar–PV, are in tune with these trends as are the smaller combustion technologies using natural gas. This modular generation also fits in with the trends in transmission and distribution of electricity, arising from the new pressures of competition in the industry, using "retail wheeling" arrangements to get power from independent power producers to individual customers. This has already occurred in the UK, where privatization and independent power production are well advanced, and indications point this way for power R&D in the USA as well. However, there is also increasing evidence that utilities are likely to reduce the timescale over which they look for returns on investment, thus precluding any development projects beyond those that will bring returns within a couple of years (IEA, 1997).

These changes give rise to the question of what are the prospects for the penetration of new energy technologies in the face of unsubsidized commercialization and intensified market competition in energy markets? It is argued by advocates of alternative energy that it is incorrect, even unfair, to demand that the costs of energy from the new technologies compete against the costs of conventional energy, when the full costs to society are not included in the prices of fossil fuels (Harrison, 1975; Roodman, 1997). The best known category of omissions are the economist's "externalities", such as the costs of property damage and health risks that are caused by pollution emissions but not factored into the cost of energy (Hill *et al.*, 1995). Some argue simply that the cost of energy in the late 1990s is too low, not only because it does not include a figure for the environmental impact of fossil fuels, but also because it does not take into consideration the scarcity value of these depletable resources (GEV, 1997). Still others argue that the economy should be considered part of the entire ecosystem and that truly sustainable development can only be achieved by doing so (Daly, 1996).

New methods of technology assessment, called the "life-cycle approach", account for all impacts, including environmental, social, macro-economic, and security factors (Curran, 1996; Sorensen, 1996). Alternative methods of pricing energy, accounting for the societal benefits of renewables as well as the costs to society of conventional energy, have also been proposed (Bernow et al., 1994; Stern & Dietz, 1997); this is known as "social costing" or "full cost pricing". Some proposals even call for broad "ecological tax reform" to make equitable adjustments for what they see as defects in the market for energy (Scheer, 1994). Given the outlook in the USA and other industrial countries, however, the adoption of any of these ideas is unlikely in the foreseeable future.

It appears unlikely that major market support will be available from governments generally. This fact, plus variabilty and uncertainty in support for R&D, make it clear that unit costs of energy from these new technologies cannot be set assuming subsidized reductions when making comparisons with the costs of conventional fuels. Until these factors change, assessment of the readiness of these alternative sources of energy must be based on a simple comparison of costs with the market prices at the time, as we have done in the preceeding chapters. Consequently, our assessments of the prospects for sustainable new technologies, as summarized in the previous chapter, must remain as disappointingly limited as they are.

In summary, politics have been and will remain a major determinant of when alternative energy technologies will be developed and on what scale when they are. The research and development process is especially vulnerable to the uncertainties and manipulations of the legislative process. It seems fairly evident that any program capable of inducing mass changes in national energy supplies within a few decades would require the political will that can only come from consensus on a national mission in any one or more of the countries having the research and industrial capacity to do so. Energy R&D funding in the USA and other industrial countries has been inadequate through the mid-1990s for any sort of assurances of energy alternatives in the long term. We can conclude here that we will be on the way to *sustainable* energy supplies only when there is a political consensus to do so. Perhaps this is all summarized by Goldhaber (1986): "the technology that gets developed is a direct result of political choices".

References

AAAS (AAAS Intersociety Working Group) (1994). *AAAS Report XIX – Research and Development FY 1995*. Washington, DC:American Association for the Advancement of Science.

Anon. (1991). *Europe Energy*, 359, June 28.

Anon. (1992). *Europe Energy*, 374, March 6.

Anon. (1993a). Basic research I, II (Editorials). *Science*, 259, 291, 579.

Anon. (1993b). Clinton's technology policy emerges (News and Comments). *Science*, 259, 1244–1245.

Anon. (1993c). Science, technology and national goals (Editorial). *Science*, 259, 743

Anon. (1993d). Industrial research, reinventing the automobile and government R&D. *Science*, 262, 172.

Anon. (1993e). Updating Vannevar Bush: academy panel calls for new strategy for science (Washington report). *Physics Today*, July, 67.

Anon. (1993f). Gore promises US leadership on sustainable development path (Environment section). *New York Times*, June 15, C4.

Anon. (1994a). US science policy, White House lauds basic research. *Science*, 265, 731–732.

Anon. (1994b). After last year's threat to NSF, Mikulsky issues another manifesto (Washington report). *Physics Today*, Oct., 57.

Anon. (1994c). Congress enacts 1995 R&D budgets, with NIST and NSF the big winners. *Physics Today*, 59–61.

Anon. (1994d). A Strategic Message from Mikulsky (News and Comments). *Science*, 263, 604.

Anon. (1994e). The Hand on Your Purse Strings (News and Comments). *Science*, 264, 192–194.

Anon. (1994f). Mikulsky Boosts NSF Budget; *Science*, 265, 469–470.

Anon. (1994g). US research spending – accounting tricks boost NSF Spending (News and Comments). *Science*, 265 (26 Aug.)

Anon. (1994h). An industry-friendly science policy – a restructuring of Britain's research councils, aimed at making academic research more useful to industry, is now taking hold – and unease is growing in some disciplines (News and Comments). *Science*, 265, 29.

Anon. (1994i). New GOP chairs size up science (News and Comments). *Science*, 266, 1796–1797.

Anon. (1994j). Science and the new Congress, Walker unveils R&D Strategy (News and Comments). *Science*, 266, 1938.

Anon. (1994k). Environmental Protection Agency – Browner to beef up outside research (News and Comments). *Science*, 265, 599.

Anon. (1995). Republican bucks basic research trend (Science Scope). *Science*, 267, 19.

Anon. (1997a). The changing energy mix. *Oil & Gas Journal*, Aug., 37–54.

Anon. (1997b). Five-year plan squeezes R&D (News & Comments section on 1998 Budget). *Science*, 276, 1328–1329.

Anon. (1997c). Friendly finish looms on spending, Congress and the budget section. *Science*, 278, 30.

Anon. (1997d). Study finds public science is the pillar of industry. *New York Times*, May 13, C1, C10.

Anon. (1997e). Competition seen threatening public purpose energy R&D. *New Technology Week*, Nov., 7.

Anon. (1998). Euphoria fades as threats emerge, US R&D budget. *Science*, 280, 819–820.

Bernow, S. *et al.* (1994). From social costing to sustainable development: beyond the economics paradigm. In: *Social Costs of Energy – Present Status and Future Trends*, eds. Hohmeyer, O. & Ottinger, R., pp. 373–404. New York: Springer-Verlag.

Blissett, M. (1972). *Politics in Science*. Boston, MA: Little Brown.

Branscomb, L.M. (ed.) (1993). *Empowering Technology – Implementing a US Strategy*. Cambridge, MA: MIT Press.

Braun, H.J. (1992). Introduction to symposium on failed innovations. *Social Studies of Science*, 22, 213–230.

Brooks, H. (1968). *The Government of Science*. Cambridge, MA: MIT Press.

Brooks, H. (1971). Can science survive the modern age? *Science*, 174, 21–30.

Brooks, H. (1993). Research universities and the social contract for science. *Empowering Technology – Implementing a US Strategy*, ed. Branscomb, L.M. Cambridge, MA: MIT Press.

Brower, M.C., Tennis, M.W., Denzler, E.W. & Kaplan, M.M. (eds.) (1993). *Powering the Midwest – Renewable Electricity for the Economy and the Environment*. Cambridge, MA: Union of Concerned Scientists.

Bush, V. (1945). *Science: The Endless Frontier, A Report to the President on a Program for Postwar Scientific Research*, Reprinted by the National Science Foundation (1990) Washington, DC.

Cassedy, E.S. & Grossman, P.Z. (1990). *Introduction to Energy – Resources, Technology and Society*, Ch. 12, Section entitled Nuclear fusion power reactors (pp. 252–262). New York: Cambridge University Press.

CCS (Carnegie Commission on Science, Technology and Government) (1994). Reprinted report of the Carnegie Commission on Science, Technology and Government. In: *AAAS Science and Technology Policy Yearbook*, Ch. 22, eds. Teich, A.H. *et al.*, pp. 211–222. Washington, DC: American Association for the Advancement of Science.

Cross, B. (1993). *European Directory of Renewable Energy Suppliers and Services*. Luxembourg: Commission of the European Community (CEC).

Curran, M.A. (1996). *Environmental Life Cycle Assessment*. New York: McGraw-Hill.

Daly, H.E. (1996). *Beyond Growth – The Economics of Sustainable Growth*. Boston, MA: Beacon Press.

Dasgupta, P. & Stoneman, P. (1987). *Economic Policy and Technological Performance*. Cambridge: Cambridge University Press.

de la Mothe, J. & Ducharme, L. (1990). *Science, Technology and Free Trade*. London: Pinter.

DOE (US Department of Energy) (1991). *National Energy Strategy 1991/1992*. Oak Ridge, TN: Office of Scientific & Technical Information.

DOE (1993). *Biofuels – Program Plan FY 1992–FY 1996*. DOE/CH10093–186, DE 93000036. Golden, CO: National Renewable Energy Laboratory.

DOE (1994). *FY 1995 Budget Highlights (Feb.)*. Washington, DC: Office of the Chief Financial Officer.

DOE (1997). *Next Generation Turbine Development*. Golden, CO: Golden Field Office. (As reported. In *Solar Today*, Sept./Oct., 41–42 (1997)).

Ehrenreich, H. (1995). Strategic curiosity: semiconductor physics in the 1950s. *Physics Today*, Jan., 28–34.

EUFORES (European Forum for Renewable Energy Sources) (1998). *The Declaration of Canaries* (Newsletter). Brussels: European Forum for Renewable Energy Sources.

Evans, L.C. (1992). Wind energy in Europe. *Solar Today*, May/June, 32–34.

Florida, R. & Kenney, M. (1990). *The Break-through Illusion.* New York: Basic Books.

Fox, M.A. (1994). The contribution of curiosity-driven research to technology. In: *AAAS Science and Technology Policy Yearbook,* eds. Teich, A.H. *et al.* Washington, DC: American Association for the Advancement of Science.

Freeman, S.D. (1974). *A Time to Choose, The Energy Policy Project of the Ford Foundation.* Cambridge, MA: Ballinger.

FSEC (Florida Solar Energy Center) (1979). *Proceedings of the Solar Energy Consumer Protection Workshop,* March, CONF–7805162, Cape Canaveral, FL: Florida Solar Energy Center.

GAO (US General Accounting Office) (1993). *Electricity Supply – Efforts Under Way to Develop Solar and Wind Energy.* Report to the Chairman, Subcommittee on Investigations & Oversight on Science, Space and Technology, US House of Representatives, April. Washington, DC: US General Accounting Office.

GEV (Greenpeace E.V. and Deutsches Institut für Wirtschaftsforschung) (1997). *The Price of Energy.* Aldershot,UK: Dartmouth Publishing.

Gever, J., Kaufman, R., Skole, D. & Vorosmarty, C. (1986). *Beyond Oil – The Threat to Food and Fuel in the Coming Decades.* Cambridge, MA: Ballinger.

Gilman, P. (1990). Some frank observations. (In: *Symposium on Plasma Physics, Public Policy, and the Future of Fusion,* (Chair Dean, S.O.)). *Physics and Society,* 19, 5–8.

Gipe, P. (1995). *Wind Comes of Age.* New York: Wiley.

Girifalco, L.A. (1991). *Dynamics of Technological Change.* New York: Van Nostrand Reinhold.

Goldhaber, M. (1986). *Reinventing Technology: Policies for Democratic Values,* London: Routledge. (Also as cited in Street, J. (1992). *Politics & Technology,* New York: The Guilford Press.)

Gore, A. (1993). *Earth in the Balance – Ecology and the Human Spirit.* New York: Plume.

Hagstrom, W.O. (1965). *The Scientific Community.* New York: Basic Books.

Harrison, D. (1975). *Who Pays for Clean Air?* Cambridge, MA: Ballinger.

Hill, R., O'Keefe, P. & Snape, C. (1995). *The Future of Energy Use,* Ch. 2, The cost of energy. New York: St Martin's Press.

Hollander, J.M. (ed) (1992). *The Energy–Environment Connection.* Washington, DC: Island Press.

IEA (International Energy Agency) (1994). *Energy Policies of IEA Countries – 1993 Review.* Paris: Organization for Economic Cooperation and Development (OECD).

IEA (1996a). *Comparing Energy Technologies.* Paris: Organization for Economic Cooperation and Development (OECD).

IEA (1996b). *The Role of IEA Governments in Energy – 1996 Update.* Paris: Organization for Economic Cooperation and Development (OECD).

IEA (1996c). *Climate Technology Initiatives – Inventory of Activities.* Paris: Organization for Economic Cooperation and Development (OECD).

IEA (1997). *Competition and New Technologies in the Electric Power Sector.* Paris: Organization for Economic Cooperation and Development (OECD).

IEEE (IEEE–USA Energy Policy Committee) (1994). *Entity Position Paper, Vision of Electric Energy Policy, July 29.* Washington, DC: Institute of Electrical & Electronic Engineers.

Kelly, K.F. (1994). A policymaker's perspective on the social responsiveness of science. In: *AAAS Science and Technology Policy Yearbook*, Ch. 17, eds. Teich, A.H., Nelson, S.D. & McEnaney, C. *et al.* Washington, DC: American Association for the Advancement of Science.

Latour, B. & Woolgar, S. (1979). *Laboratory Life – The Social Construction of Scientific Facts.* Beverly Hills, CA: Sage.

Lyons, P.K. (1994). *Energy Policies of the European Union.* London, UK: EC Inform.

Maglich, B. (1991). Can the nation afford not to pursue research on aneutronic nuclear power? *Physics and Society*, 20, 4–6; Holt, R. (1991). Tokamak magnetic fusion and the first generation of fusion power. *Physics and Society*, 20, 6–7. A pair of articles presenting the arguments for and against maverick proposals for research on reactors utilizing nuclear-fusion reactions not involving neutrons. Letters following on the same subject: *Physics and Society*, 20, Oct.

Mulkay, M.J. (1975). Norms and idealogy. *Science, Social Science Information*, 15, 637–656.

NAS (National Academy of Science, National Academy of Engineering & Institute of Medicine) (1992). *The Government Role in Civilian Technology – Building a New Alliance.* Washington, DC: National Academy Press.

Nicklas, M. (1997). 40% renewable energy by 2020: an ISES goal, a global necessity. In: *Advances in Solar Energy*, Vol. 11, Ch. 8, ed. Boer, K.W., pp. 415–459. Boulder, CO: American Solar Energy Society.

NYSERDA (New York State Energy Research & Development Authority) (1997). *Toward the 21st Century, A Three–Year Research Plan for New York's Energy, Economic and Environmental Future: 1997–2000.* Albany, NY: New York State Energy Research & Development Authority.

OECD (1994). OECD Statistics on the Member Countries. *OECD Observer*, 188 (Suppl.), June/July.

Pye, M. & Nadel, S. (1997). Energy technology innovation at the state level: review of state energy RD&D programs. Washington, DC: American Council for an Energy–Efficient Economy.

Ray, G.F. & Uhlmann, L. (1979). *The Innovation Process in the Energy Industry.* Cambridge, UK: Cambridge University Press.

Remington, J.A. (1988). Beyond big science in America: the binding of inquiry. *Social Studies of Science*, 18, 45–72.

Rogers, E.M. & Shoemaker, F.F. (1971). *Communication of Innovation: A Cross-cultural Approach*, 2nd edn. New York: The Free Press.

Roodman, D.M. (1997). *Getting the Signals Right: Tax Reform to Protect the Environment and the Economy.* Washington, DC: World Watch Institute.

Rosen, L. & Glasser, R. (1992). *Climate Change and Energy.* New York: American Institute of Physics.

Rosenberg, N., Landau, R. & Mowery, D.C. (1992). *Technology and the Wealth of Nations.* Stanford, CA: Stanford University Press.

Sarewitz, D. (1996). *Frontiers of Illusion – Science, Technology, and the Politics of Progress.* Philadelphia, PA: Temple University Press.

Scheer, H. (1994). The economy of solar energy. In: *Advances in Solar Energy,* Vol. 9, Ch. 6, ed. Boer, K.W., pp. 307–338. Boulder, CO: American Solar Energy Society.

Schell, D.J., McMillan, J.D., Philippidis, G.D., Hinman, N.D. & Riley C. (1992). Ethanol from lignocellulosic biomass. In: *Advances in Solar Energy,* Vol. 7, Ch. 10, ed. Boer, K.W., pp. 373–448. Boulder, CO: American Solar Energy Society.

Schmitt, R.W. (1994). Public support of science: searching for harmony. *Physics Today,* Jan, 29–33.

Science, Special Issue (1993). Science in Europe ”93. *Science,* 260, 1733–1758.

Smith, B. (1990). *American Science Policy since World War II.* Washington, DC: The Brookings Institution.

Sorensen, B. (1996). Life cycle approach to assessing environmental and social externality costs. In: *Comparing Energy Technologies,* Ch. 5, pp. 297–331. Paris, France: International Energy Agency, OECD.

SRCC (Solar Rating Certification Corporation) (1993–4). *Operating Guidelines for Certification of Solar Collectors,* SRCC-OG-100, 1994 (update); *Operating Guidelines and Minimum Standards for Solar Hot Water Systems,* SRCC-OG-300, 1993 (under revision). Washington, DC: Solar Rating Certification Corporation (an industry Group).

Stern, P.C. & Dietz, T. (1997). Consumption and sustainable development (Letter). *Science,* 276, June. (In response to Policy Forum of 4 April, 1997.)

Stobough, R. & Yergin, D. (eds.) (1983). *Energy Future.* New York: Vintage.

Teich, A.H., Nelson, S.D. & McEnaney, C. (eds.) (1994–5). *AAAS Science and Technology Policy Yearbook.* Washington, DC: American Association for the Advancement of Science.

Tornatzky, L.G. & Fleischer, M. (1990). *The Processes of Technological Innovation.* Lexington, MA: Lexington Books.

van Vort, W.D. & George, R.S. (1997). Impact of the California clean air act. *International Journal of Hydrogen Energy,* 22, 31–38.

Weinberg, C.J. (1994). The electric utility; restructuring and solar technologies. In: *Advances in Solar Energy,* Vol. 9, Ch. 5, ed. Boer, K.W., pp. 269–306. Boulder, CO: American Solar Energy Society.

Williams, S. & Bateman B.G. (1995). *Power Plays – Profiles of America's Independent Renewable Electricity Developers.* Washington, DC: Investor Responsibility Research Center.

Yanarella, E.J. & Green, W.C. (1987). *The Unfulfilled. Promise of Synthetic Fuels – Technological Failure, Policy Immobilization or Commercial Illusion.* New York, NY: Greenwood Press,

Further reading

Anon. (1994a). Counting what counts: 1995 budget skimps science, boosts technology (Washington Report). *Physics Today,* 49–55.

Anon. (1994b). O'Leary stands up for DOE (Science Scope). *Science,* 266, 1795.

Anon. (1994c). 1995 Science budget – hitting the President's target is mixed blessing for agencies (News and Comments). *Science,* 266, (14 Oct.).

Anon. (1994d). NSF gears up for a building boom – campaign to use peer review, not pork, to rebuild new labs (News and Comments). *Science,* 265, 1516–1518.

Anon. (1994e). EPA plans to squeeze pork 'til it squeals (Science Scope). *Science,* 265, 1999.

Anon. (1994f). Academic earmarks – pork takes toll on research projects (News and Comments). *Science,* 265, 2004.

Anon. (1994g). Failing peer review, school gets pork funds (Science Scope). *Science,* 266, 19.

Anon. (1994h). The politics of alternative medicine – . . . a powerful senator and advocates of unconventional medicine are shaping the . . . research agenda (News and Comments). *Science,* 265, 2000–2002.

Anon. (1995). Energy laboratories, report to stress research over close ties to industry. *Science,* 267, 446–447.

ASEA (American Solar Energy Association) (1994). Chairman's Corner (Editorial comments representing the viewpoint of the ASES). *Solar Today.*

Boring, E.G. (1952). The validation of scientific belief. *Proceedings of the American Philosophical Society,* 96, p. 535.

Brower, M. (1992). *Cool Energy,* Appendix B, *US Renewable Energy Funding – 1974 to 1992.* Cambridge, MA: MIT Press.

Cohen, I.B. (1952). Orthodoxy and Scientific Progress. *Proceedings of the American Philosophical Society,* 96, 509–510.

Cohen, L.R. & Noll, R.G. (1991). *The Technology Porkbarrel.* Washington, DC: The Brookings Institution.

Colitti, M. (1994). Economic stagnation and sustainable development. In: *Sustainable Development and the Energy Industries,* ed. Steen, N., pp. 35–43. London: The Royal Institute of International Affairs, published by Earthscan.

Durning, A.T. (1994). *Redesigning the Forest Economy,* Ch 2, State of the World – 1994, pp. 22–40. Washington, DC: World Watch Institute.

ETSU (Energy Technology Support Unit) (1994). *An Assessment of Renewable Energy for the UK.* Harwell, UK: UK Renewable Energy Programme, Department of Trade & Industry.

Hart, D.M. & Victor, D.G. (1993). Elites and the making of US policy for climate change. *Social Studies of Science,* 23, 643–680.

Hohmeyer, O. & Ottinger, R. (eds.) (1994). From social costing to sustainable development: beyond the economics paradigm. *Social Costs of Energy - Present Status and Future Trends,* pp. 373-404. New York: Springer-Verlag.

Landau, R. & Rosenberg, N. (1986). *The Positive Sum Strategy.* Washington, DC: National Academy Press.

Merton, R.K. (1965). The ambivalence of scientists. In: *Science and Society,* ed. Kaplan, N. Chicago, IL: Rand McNally.

Mulkay, M.J. (1976). The mediating role of the scientific elite. *Social Studies of Science,* 6, 445–470.

Peterson, J. & Mastaitis, V. (1993). Target 2000: renewable energy in New York state. *Solar Today,* July/Aug.

Polanyi, M. (1964). *Science, Faith and Society.* Chicago, IL: University of Chicago Press.

Science, Special Issue (1992). Science in Japan: perspectives. *Science*, 23 Oct.: (a) Hirano, Y. public and private support of research, 582–3; (b) Arima, A. Underfunding of basic science in Japan, 590–1.

SEIA (Solar Energy Industries Association) (1994). The solar commercialization challenge – the SEIA's FY 1995 appropriations recommendations for the US Department of Energy. *Solar Industry Journal*, First Quarter, 1994.

Veziroglu, T.N. (1997). Hydrogen movement and the next action, fossil fuels industry and sustainable economics. *International Journal of Hydrogen Energy*, 22, 555–556.

Vories, R. & Strong, H. (1988). *Solar Market Studies: Review and Comment.* Golden, CO: Solar Research Institute.

WEC/IIASA (World Energy Council/International Institute for Applied Systems Analysis) (1995). *Global Energy Perspectives to 2050 and Beyond, Report 1995.* London, UK: World Energy Council.

Ziman, J.M. (1968). *Public Knowledge.* Cambridge, UK: Cambridge University Press.

Appendix A Energy cost analysis

The basis for cost comparisons used throughout this book is the levelized unit cost of energy (Marsh, 1980; Hock *et al.*, 1992; de Laquil *et al.*, 1993). In this calculation of cost of an energy source, the sum of annual fixed costs (mostly capital costs) and annual variable (operating) costs is divided by the annual energy output of the source to give the (average) cost per *unit* of energy output. The annual fixed costs are assumed to be distributed uniformly ("levelized") over the depreciation lifetime of the equipment and include the interest or equity return required from the initial investment. Variable costs, such as fuel costs and other operational costs, are also assumed to be levelized according to the time value of money over the years. As applied to conventionally generated electricity, cost of energy is often limited to the power plants and does not then include the costs of transmission and distribution. As such, they are said to be the "busbar" costs of the generating plant, meaning that they stop at the interconnection point of the conductors to the outside.

This method of representing (unit) energy costs is standard in the utility industries for both electric energy ($/kW-hr) and heat energy ($/MBTU) (Marsh, 1980; Stoll, 1989; Hock, *et al.*, 1992). It affords direct comparisons between the costs of competing new energy sources trying to penetrate the market and direct comparisons with prices of existing sources in energy markets. Other methods of energy cost analysis are in use, such as life-cycle payback (not to be confused with *environmental* life-cycle cost) and discounted cashflow analyses, but these are more tailored to individual financial decision making (Wiser, 1997) rather than broad market comparisons of the type made here. We have, nonetheless, cited two examples of payback analysis for solar technology in Chapter 3, one for a homeowner and another for an industrialist, each making the decision whether to invest in solar heat collectors.

The outcomes from these various methods will, of course, be similar: one would not expect a favorable payback, for example, from a technology that clearly has a non-competitive unit cost in its energy market. Payback analysis, however, may be able to include regional or other situational factors that fit a particular case where the investment decision is being made with marginally competitive conditions and discern an advantage for investment. However, any attempt to project a series of cash flows for items such as fuel savings will inevitably suffer from the uncertainty of future prices. In any case, there is no simple way to apply a collection of such individual payback analyses to an entire energy market, and most studies of the new technologies, therefore, use the levelized cost of energy.

The levelized unit cost of energy (LCOE) may be expressed more precisely in the following form:

LCOE = LRR/Annual energy output

where LRR is the levelized (annual) required revenue.

LRR = LCC + LAE + O&M

Where LCC is the levelized cost of capital (initial capital cost multiplied by the fixed-charge rate (FCR)); LAE is the levelized annual operating expenses (including fuels and variable operation and maintenance charges); and O&M is the fixed operation and maintenance charges.

The initial capital cost is simply the investment cost, as laid out at the time of construction or installation of the energy plant. The FCR is a standard in the utility industry to include depreciation, (bond) interest, return on equity, taxes (both income and *ad valorum* types), and tax credits. The FCR is customarily levelized using a discount rate, either market derived for private sector projects or a social discount rate for public projects (Marsh, 1980; Stoll, 1989).

If taxes are not included, as they might not be for public projects or stand-alone projects, the FCR reduces to the capital recovery factor (CRF):

$$CRF = i(1 + i)^n / (1 + i)^n - 1$$

used in standard financial formulae, where i is the annual interest rate or rate of return and n is the depreciation lifetime (in years). Values of the CRF, for given rates of return and lifetimes, may be found in tables (Park & Jackson, 1984) or calculated directly. The simple CRF has been used in this volume to calculate costs of energy unless otherwise noted. In some of the exceptions, such as solar–thermal generation, the difference from the FCR has resulted only from the tax credits. In any of the grid-connected applications, including wind and solar generation, the full FCR will usually be required to make optimal integration with conventional generation (Hock *et al.*, 1992; Kelly & Weinberg, 1993).

There are methods of energy cost analysis that account for more than simply the time value of money, such as the leveling or time discounting we have been considering here. One of these is the capital assets pricing model (CAPM), which accounts for uncertainty of outcomes in costs and technological change (Awerbuch, 1995). This approach explores a means of levelizing that attempts to account for the "experience-curve" drops in production costs (for PV cells) with increasing mass production. This approach, still in the research stage, could lead to lower cost estimates for new technologies, leading to earlier adoption in the market.

References

Awerbuch, S. (1995). New economic perspectives for valuing solar technologies. In: *Advances in Solar Energy,* Vol. 10, Ch. 1, ed. Boer, K.W. Boulder, CO: American Solar Energy Society.

de Laquil III, P., Kearney, D., Geyer, M. & Diver, D. (1993). Solar-thermal electric technology. In: *Renewable Energy – Sources for Fuels and Electricity*, Ch. 5, eds. Johansson, T.B., Kelly, H., Reddy, A.K.N. & Williams, R.H. Washington, DC: Island Press.

Hock. S., Thresher, R. & Williams, T. (1992). The future of utility–scale wind power. In: *Advances in Solar Energy,* Vol. 7, Ch. 9, ed. Boer, K.W., pp. 309–371. Boulder, CO: American Solar Energy Society.

Kelly, H. & Weinberg, C.J. (1993). Utility strategies for using renewables. In: *Renewable Energy – Sources for Fuels & Electricity,* Ch. 23, eds. Johansson, T.B., Kelly, H., Reddy, A.K.N. & Williams, R.H., pp. 1011–1069. Washington, DC: Island Press.

Marsh, W.D. (1980). *Economics of Electric Power Generation.* Oxford: Oxford University Press.

Park, W.R., Jackson, D.E. (1984). *Cost Engineering Analysis*, 2nd edn. New York: Wiley.

Stoll, H.G. (1989). *Least Cost Electric Utility Planning.* New York: Wiley.

Thurman, A. (1984). *Fundamentals of Energy Engineering.* New York: Prentice Hall.

Wiser, R.H. (1997). Renewable energy finance and project ownership. *Energy Policy*, 25, 15–27.

Appendix B **Glossary**

BTU	British thermal unit (heat required to raise 1 lb water 1 °F)
cal	calorie (heat required to raise 1 g water 1 °C)
CRF	capital recovery factor (see Appendix A)
CF	capacity factor (see Terms, below)
DHW	domestic hot water
DSM	demand side management
EJ	exajoule (10^{18} joules (energy))
EV	electric vehicle
FCR	fixed-charge rate (see Appendix A)
FCV	fuel-cell vehicle
GJ	gigajoule (10^9 (billion) joules (energy))
GW	gigawatt (10^9 (billion) watts (power))
ha	hectare (10 000 square meters land area)
HHV	higher heating value (for fuels with heat of vaporization)
HV	heating value
HVAC	high voltage alternating current
HVDC	high voltage direct current
Hz	hertz (cycles per second)
ICE	internal combustion engine
IPH	industrial process heat
K	Kelvin (absolute temperature scale, metric system)
kcal	kilocalorie (1000 calories)
km	kilometer (1000 meters)
kW	kilowatt (1000 watts (power))
kW_e	kilowatt of electric power output capacity
kW_p	kilowatt peak electric power output (for solar PV units)
kW-hr	kilowatt hour (kilowatts x hours (electric energy))
LDC	less-developed country
LHV	lower heating value (for fuels with heat of vaporization)
LRR	levelized revenue requirement (see Appendix A)
MBTU	million BTU (10^6 BTU (thermal energy))
MW	megawatt (10^6 (million) (power))
MW_e	megawatt of electric output capacity
MW_p	megawatt peak electric output (for solar PV installations)
O&M	operation and maintenance
PV	photovoltaic
QUAD	quadrillion BTU (10^{15} BTU (primary energy measure))
r.p.m.	revolutions per minute (rotational motion)
SDHW	solar domestic hot water

TW	terawatt (10^{12} trillion watts (power))
W_p	watts peak electric power output (for PV cells)

Prefixes

kilo (k)	one thousand (10^3)
mega (M)	one million (10^6)
Giga (G)	one billion (10^9)
Tera (T)	one trillion (10^{12})
quad	quadrillion (10^{15})
exa (E)	10^{18}

Subscripts

e	electric power output capacity
p	peak electric output
th	thermal output

Conversion of units

British units to metric units

Energy

1 BTU = 1055 J = 252 cal
1 MBTU = 1.055 GJ = 1055 kJ
1 QUAD = 1.055 EJ = 1055 x 10^9 MJ = 10^{15} BTU
3412 BTU = 1 kW-hr = 3600 kJ

Distance and speed

1 inch = 2.54 cm
1 ft = 0.3048 m
1 mile = 1.609 km = 1609 m
1 mph = 1.609 km/h = 0.447 m/s

Area

1 ft^2 = 0.0929 m^2
1 acre = 0.4051 ha

Volume

1 ft^3 = 0.0283 m^3
0.833 (UK) gallon = 1 (US) gallon = 3.785 liters
1 barrel = 42 gallon (US) = 159 liters

1 acre-ft = 43 560 ft³ = 1228 m³

Mass

1 lb = 0.4536 kg
1 (short) ton = 2000 lb = 0.9072 metric tonne (mt) = 907.2 kg

Mixed measures

1 ton/acre = 2.24 mt/ha
BTU/SCF = 37.3 GJ/m³
1 MBTU/ton = 1.16 GJ/mt
$/MBTU = $ 0.945/GJ (= 0.756 ecu/GJ, when $1 = 0.8 ecu
(where ecu is the European currency unit))

Terms

absorption cooling	an air-conditioning cycle that operates on absorption and vaporization of the working fluid (e.g. ammonia) with another fluid (e.g. water)
acre	43 560 ft² land area
biomass	non-fossil organic (plant or animal) material
base load	the minimum portion of the electric load demand that is present all or most of the 8760 (365 x 24) hours of the year.
busbar costs	costs of electric generation incurred at the generation station only, not including costs of transmission or distribution
capacity factor	ratio of annual average power output to rated output or percent annual hours of equivalent full-rated output
condenser	a heat exchanger that cools a gaseous coolant, causing it to condense to the liquid state, thereby giving up heat to be carried out to the surroundings
coolant	the circulating fluid or gas that transfers heat from its source to its sink in a heating or cooling system
emittance	the fraction of absorbed heat that is reradiated; this varies from one material to another
fixed charges	costs that do not vary with output, such as capital charges (see Appendix A)

greenhouse gases	gases that contribute to global warming by absorbing infrared radiation from the earth, including carbon dioxide, nitrous oxide, methane and chlorofluorocarbons
grid	the interconnected network of transmission and distribution lines of an electric power system
grid support	supplementary generators connected at points on the network to help to sustain voltage levels and supply particular loads
(head) hydraulic head	the height from which water must fall to deliver its energy for hydroelectricity generation
herbaceous	non-woody plants
hydrolysis	the conversion of a complex organic compound, such as cellulose, into a simpler organic compound, such as a sugar, by reaction with water and enzymes
insolation	loosely, the amount of sunlight received; more strictly, the intensity of incident sunlight in terms of power per unit of area, such as watts/m^2
installed capacity	the aggregate of the rated power output capacities of all generators on a power system (MW_e or GW_e)
joule	the basic energy unit of the metric international system
load factor	the ratio of average power demand to peak load demand on an electric power (grid) system
metric ton (mt)	tonne (1000 kg)
peak load	the maximum portion of the electric load demand, present relatively few hours out of the year
p–n junction	the thin separation in a semiconductor between a region with positive-charged carriers ("holes") and that with negatively charged carriers (electrons)
power factor	the ratio of watts (real power) to volt-amperes (apparent power) of an electric load
primary energy	energy derived directly from a primary resource of energy, such as fossil fuels, nuclear sources, sunlight, wind, or hydro sources, which may be converted

	into a secondary source of energy, such as electricity or hydrogen. The conversion process can never convert 100% of the primary energy
pumped storage	a hydroelectric plant that stores energy by having water pumped to a reservoir at a higher elevation and then recovers the energy by releasing the water and converting the energy by hydro generation
prime mover	the machine, such as a turbine, that supplies the mechanical drive for an electric generator
run of the river	a hydroelectric plant that operates from the flow of a river, without the pressure of a head or the power generated from the average flow of a river
stand alone	operation of a generator not connected to the grid
spinning reserves	the aggregate of all generators on a power system running, but not supplying power, and available to supply power if needed. Part of operation required for reliability

Appendix C **The conduct and management of research and development**

Having described *what* the energy R&D establishment is in Chapter 11, we can also look at *how* it functions. For the purposes of this book, we are most interested in how this establishment functions at the working level, since it is there that we can get insight into the prospects for breakthroughs and innovations for prospective technologies. We need to identify those attributes in the establishment that are most determinant of high creativity. Conditions where creativity and inventiveness are promoted are critical throughout the R&D process, starting in basic research and carrying through applied research and development. Conversely, we are interested in those conditions that discourage or impede free thinking and creativity. At the same time, we are interested in where – that is, along what paths – creativity can lead.

II.1 **Research**

Research is a human endeavor, influenced as much by social behavior as by the facts of nature. It is well for public policy makers and business decision makers to delve, at least briefly here, into the institutions of research in order to gain some understanding of what outcomes can reasonably be expected from the research process. The conditions conducive to high creativity in basic research revolve around the discipline of inquiry and the circle or groupings of those working in a field. These circles are known to center on the locale where the work is carried out such as the university or research lab, and the circles in which research results are reported and discussed.

High prestige is attached to a university or laboratory where notable work has been done, status becoming most prestigious if Nobel Laureates are in residence (Zuckerman, 1977). Young scientists are attracted to these institutions for the stimulating atmospheres and to further their own careers. One has only to recall the stories told of the big discoveries in science (Fermi, 1954; Watson, 1981) to sense the atmospheres of intellectual challenge and free-ranging thinking that held sway in the centers where these advances were made. The locales in these stories were universities, such as Chicago, and laboratories, such as Cold Spring Harbor, where high creativity was exhibited in scientific disciplines such as nuclear physics and molecular biology. Of course, the atmosphere of excitement and stimulation was created by the brilliant individuals (Hoffman, 1972;

Gleich, 1993) at the center of activity, but those institutions had the vision to provide the physical environment.

It should be recognized, however, that such ideal atmospheres, with free-ranging ideas, do not always exist within the institutions or collegial circles of a given scientific specialty. A "group think" mentality can grow and dominate the range of new ideas that are pursued. Scientific research has been observed to operate within its established institutions by behavior characterized as: ". . . a corporate activity maintained by social processes . . ." (Blisset, 1972; Richter, 1972; Zuckerman, 1977). Consequently, the institutions of scientific and academic research have been said to operate very much on a "consensual" basis, enforcing certain norms of conducting research and tending to set research goals that seek "originality within conformity" (Mulkay, 1991). Indeed, most scientists would contend that it does so out of necessity in order to preserve the integrity of science (Crease & Samios, 1989). As such, new ideas must be "certified" (Storer, 1966). However, critics would question the objectivity of the "authorities" in many cases (Sarewitz, 1996).

Through the various professional societies and informal networks, effective consensual control is exercised over the awarding of research funds, the selection of papers for publication, and even the conferring of honors (Polanyi, 1951). The lines of inquiry are tightly tied to the groupings of the various research specialties (Latour & Woolgar, 1979). While it has been debated as to whether social groupings in science have a causal relation to the lines of inquiry pursued or vice versa, there seems little doubt that the groupings do exist and that power is exercised over the choice of paths from these groups in the various areas of research.

These social institutions of science are thought to have their roots in the research universities (Blissett, 1972; Richter, 1972) . The socialization that akes place in the training and initiation of young scientists to research can lead to a conformity of approaches under the tutelage of mentors, including the choices of research problems. Informal cliques can form at institutions or professional societies (GRC, 1973) that can wield powers of decision on support of research projects or the acceptance of publications. In these situations, which may develop in a given scientific field, mavericks are not gladly suffered (Maglich, 1991) and those established in the field can react unfavorably to unorthodoxy. Indeed, the notion of "unfettered" research is said to be a myth by critics of the scientific establishment, who claim that the decision-making processes of the scientific establishment are not accountable to outside bodies such as the public (Sarewitz, 1996).

The peer review process is where cliques or the "old boy's network" might be popularly thought to be at work (Snow, 1951, 1964), excluding people for a variety of reasons including class, race, or gender. The first area where one might suspect biases to be present would be in the awarding of research funding. This popular view of biases, however, did not seem to be very evident in an audit of peer review of federal agencies, although some degree of difference in awards related to gender were evident (Marshall, 1994). The audit did find, however, an influence on the ratings given to research proposals that correlated with the personal familiarity of the reviewer with the applicant, which would appear to correspond to earlier findings in behavioral studies of the "scientific elites" (Blissett, 1972; Mulkay, 1976; de la Mothe & Ducharme, 1990). In this way, a subtle unspoken mode of selection and patronship is often at work, grounded in notions of meritocracy and in mutual esteem amongst recognized members of the elite circles of the various fields of science and (academic) engineering. This behavior has, in the past, led to charges of an unwarranted concentration of research support in the "elite" institutions in particular sections of the country (Anon., 1991, 1994).

Following the funding of research projects, the path that the research inquiry follows will usually comply with established norms and the project plan. The final point in the course of a given project at which such norms can be enforced is in the peer review of publication in scientific journals and books. The universal expectation is that every journal paper is published only after it has been reviewed by reputable scientists in the field, a reasonable requirement in itself. Here too, however, the operation by consensus is customarily in force and tends to favor continuation of established lines of research (Blissett, 1972).

The origins of these circles, and their codes of behavior, are thought to be in the "process of socialization" that takes place in the education and research training of doctoral students (Blissett, 1972; Richter, 1972). The most ambitious students typically go to great lengths to find mentors at the forefront of their fields. Once with a mentor, they follow his/her example of how to conduct first-rate research and are subject to judgments, often severe, not only on the quality of their work but also on the directions of their inquiries. According to behavioral studies (Zuckerman, 1977): ". . . the least important aspect of their apprenticeship is acquiring substantive knowledge from their masters . . ." and the most important was ". . . the contact: seeing how they operate . . .".

The entire process of initiation to research is overlaid with the traditions and social norms of the mentor and his field (Roe, 1953), which influence the conducting of observations and making of predictions, even under the scientific maxim of "accepting whatever nature's answers are. . .". The mentors themselves have probably gone through this "stringent socialization" and will ". . . always know a good problem when they come upon it . . ." (Richter, 1972). That is, they will normally select research projects for their graduate students that are similar to those of others working in the same specialty (Mulkay, 1976) and, therefore, the most likely to extend results leading to the consensual goals of the field.

Once the young researcher is launched into his/her career, the choice of a research topic is a fateful one (Brooks, 1968). Professional success will only come with recognition and funding, which will most likely accompany those problems and research techniques falling within the current consensus of the established researchers in the field. Furthermore, once established, the researcher is most likely to get renewal of funding by continuing already established lines of inquiry. The reasons for this policy by the funding agencies is not simple conservatism but rather notions of efficiency and financial prudence. That is, already established areas are the most ready to be fruitful in producing results, even if they do not promote novel approaches and outcomes. The consequences of such funding policies, therefore, is to continue particular research approaches, making them into a "tradition", while excluding others that might be unusual.

In this way, we can see that social influences may compete with the objective facts revealed by experimentation to influence the subsequent research path. The ideal concept of science and research is that progress and new understanding proceed purely as human beings are able to perceive and deduce the facts of nature. Behavioral studies indicate that most natural scientists give credence to this ideal and function in their work by norms that adhere to such beliefs. For example, social scientists and historians of science have observed that the physical-science literature uniformly makes no reference to the social factors or human processes involved in producing the theories or results that are reported, to the extent that "Reading scientific treatises is like moving through the observations of disembodied intellects . . ." (Blissett, 1972). The traditions for such ways of thinking and communicating go back to the earliest days of European science.

None of this is intended here to imply that the "corporate" establishment of science should not be accorded public support, however. The social behavior reviewed here is an imbedded part of our culture; indeed, it is the product of a history going back to the origins of Western science. The purpose of the inclusion of these behavioral patterns here has been to give an insight for policy makers who may themselves not have a sense of the world of science and research and what could reasonably be expected from it. It would be well, for example, for the energy planner or decision maker not to assume that the progress of research is simply a linear process following the paths of nature without human diversion. Furthermore, it would be well not to assume that research – least of all basic research – can be usefully channeled into certain paths. What *is* well to assume, is that the process of basic research, with all of its insularity and uncertainties, will ultimately produce outcomes underpinning future technological advance. At the same time, however, the public has a right to expect accountability if it is going to support undirected research.

II.2   Development

Turning now to the next stage of R&D, in the development phase there is no difficulty in identifying the paths of projects since they are defined by specific technological objectives. Much the same is true of applied research, although the objectives there will be more broadly defined. Usually, applied projects are scheduled along a time line to completion. While such schedules frequently show slippage along the course to completion, forecasting the date of completion for an individual project is usually straightforward. However, as an entire program is forecast – such as for the solar-thermal or biomass fuels programs – more uncertainty creeps in. Delays from individual projects can accumulate, often with interrelated aspects causing delays in progress overall. Finally, unforeseen problems or limitations of either a technical or an economic nature can slow progress in a development program, such as we have seen in photovoltaics and batteries. This may be a consequence of failure to achieve basic results, requiring more research with an uncertain timescale.

A number of highly creative development projects have been described, most notably within the field of computer R&D (Kidder, 1981; Freiberger, 1984; Reid, 1984; Rogers & Larsen, 1984). Development projects often can have very different driving forces to those operating in scientific research. For example, in computer R&D, development engineers are often working in laboratories of private companies engaged in highly competitive markets for new computers and software. More specifically, the archetype locale for computer development is a small, newly started company with young, highly motivated engineers who are "working 90 hours a week and loving it" (Florida & Kenney, 1990). The motivation for these "high-tech think workers" is not just the promise of profit shares but also the thrill of invention for young people who feel that ". . . the difference between tools and toys is not much".

The small venture-capital companies that have engaged in such feverish development have made large commitments to R&D (much more "D" than "R") in their start-up periods, often several times their annual sales volume. Their R&D is highly focused on the final commercial product and their organization is such that their engineers are kept aware of all other company activities that lead to that product. There is typically, in this regard, an open-door policy between the management and the engineers to air new ideas, and managers are even encouraged to "walk around" the company's labs to maintain cross-communication in the course of these intense efforts. Informal social events, such as Friday afternoon "beer busts", are provided to help the informal exchange of ideas.

All of these policies create a culture of free-wheeling innovation that has been characteristic of the computer industry (Florida & Kenney, 1990). The highly motivated engineers respond enthusiastically to the working culture in these small venture companies. This is especially so when they compare it with that of the larger companies, which have formal management structures, simply passing decisions down from the top. These young engineers also value being able to see the end-product of their inventive efforts, something they are less likely to do in the large firm.

It should be recognized, however, that these small venture-capital operations are high risk in nature and relatively impermanent. For every successful start-up there might be a dozen failures. Frequent buyouts and the high-mobility turnover of the high-tech specialists all contribute to a very transient aspect to this "breakthrough economy". Hypercompetition leads to frequent changes in suppliers and a high rate of broken contracts. Product quality can suffer when the competitiveness centers on price alone. Very little technical information is exchanged between these companies, except that taken by "stolen" employees (Florida & Kenney, 1990). Also, public relations campaigns to promote buyouts or for going public can create inflated expectations for the new products on the outside and poor morale inside, an aspect that might well have been applicable at various points in the history of PV development.

A breakthrough-economy atmosphere of this type has not been widespread in the energy R&D establishment, although the potential for it exists. One indication of this potential was the furor over "cold fusion" (see Chapter 8). Whereas there was intense interest in the improbable physics of the claims by the proposers, it was the prospect of superabundant energy from such a small, simple device that made it much more than a scientific curiosity. Such prospects bring to mind the original claims for nuclear (fission) technology or the current claims for solar–hydrogen, neither of which presently carry the credibility to stimulate sufficient support, public or private, to mount breakthrough-economy efforts in energy R&D. Nonetheless, some lessons for spurring innovation may be learned from the very successful modes of R&D operation of high-tech industries such as computers, communications, and biotechnology.

Danish wind turbines, as mentioned earlier, provide an example of innovative development in energy R&D (Gipe, 1995). This new generation of machines was developed by private companies, having tax incentives as the only form of government support for prospective purchasers. The design strategy was pitched to the market, which was electricity supplies to farms in the Netherlands. With this market in mind, these companies developed small machines of simple and robust constructions. These designs were able to deliver electricity at a lower cost than previous models (including American models). Their costs proved competitive with the prices of conventionally generated electricity, especially in the good wind areas of the north coast of Europe.

In summary, the conduct of research and development is seen as a complex and uncertain process, with social and institutional aspects playing significant roles in the courses taken and the outcomes. The conditions for creative efforts are recognizable, as are those that serve to limit those efforts. The recognition of these conditions should contribute to the formulation of better policy for energy R&D and improve the administration and management of R&D projects. The management of the processes of technological innovation, however, involves other behavioral aspects, economic as well as social, and will be discussed in the next section.

II.3   Technological change and the management of R&D

Policy decisions regarding the support of any R&D programs, including those for energy, should take account of what is known about the process of technological innovation and the effects of structure and organization of research efforts on progress. This applies in particular to industrial R&D. A number of factors have been considered for technological forecasting of the outcomes of industrial R&D and can, therefore, be considered in policy decisions for support. The size of firms, for instance, has received a good deal of attention in the past (Mansfield, 1968; Baldwin & Scott, 1987). The theory has been that large firms are more likely to make significant investments in R&D in view of their better ability to absorb the risks and uncertainties of the markets for new or innovative products. In addition, these theories suggest that the profitability of innovations will depend on the marginal costs versus the volume of the new production; this also seems to favor large firms. Furthermore, at the other end of the scale, there is an evident threshold size of small firm necessary to justify an R&D effort.

While such theories have been challenged, it is a fact currently that a few large firms, such as General Motors, dominate investments in the USA in industrial R&D. Four firms account for 20% and 100 firms account for 79% of R&D expenditures nationwide (Tornatzky & Fleischer, 1990). Despite the concentration of investments in R&D, however, the outcomes in terms of inventions and new products show that the large firms produce only about half of the innovations nationwide, while small firms produce a disproportionate one third. Indeed, the "breakthrough economy" in computer development demonstrates the strong innovative capabilities of the small firm in that industry. This does not hold true for Japan, however, where large firms produce about 80% of all innovations, including those for computers.

It has been suggested that the economic characteristics of an industry, including the concentration of firms, the barriers to entry of a firm into the market, and the market structure, have determined rates of innovation within that industry (Baldwin & Scott, 1987). Competition for a given market has been shown to stimulate R&D investments in some cases and inhibit it in others, depending on the market structure, the size of the firm relative to that market, and the R&D strategy of the firm (Tornatzky & Fleischer, 1990). High-technology firms and firms in technologically developing industries depend more on sales from new products and, therefore, invest proportionally more on R&D than other types of industry. This was probably the case with the Danish wind turbine manufacturers mentioned above. These firms have been called the "high-impact" firms: they have a high product-failure rate but are, nonetheless, successful overall.

At the project level, management and the roles of individuals become key determinants of innovation within any given company. Within development projects, as in any functioning organization, people fall into particular roles, not all defined by the formal structure. Successful R&D projects have been observed to have "gatekeepers" or "boundary spanners", who serve as ad hoc communicators outside and inside the project group (Tornatzky & Fleischer, 1990). They keep abreast of technological developments elsewhere and often are aware of customer needs in the potential market for the group's R&D outcomes. Other informal roles are filled as well, such as the "product champion" who sells the group's new ideas to management to get further support. Then there is usually a "problem solver" (again without formal title), who is the one who consistently comes up with solutions to technical problems or refines earlier technical ideas.

Regardless of how well R&D projects function internally, however, an important factor in determining their productive worth in the private sector is their connection to corporate planning. In the past, it was usual to exclude R&D personnel from strategic planning with top management. Under such an arrangement, the management's tendencies were to favor short-range R&D and, often, pet projects. With the influence of people from the technological frontier, however, strategic planning is more likely to have a mix of long-range with short-range and higher-risk with low-risk projects – all of which leads to more innovative efforts. Here we see the links with the small-firm, technology-breakthrough mode of operation described in the previous section.

Overall, one the most important factors that has been identified for successful R&D efforts is *communication* (Tornatzky & Fleischer, 1990). Clarity in communicating project needs and responsibilities, for example, has been found to be a key factor influencing the effectiveness of the R&D group. Clarity on the expected performance requirements of the new technology is especially important in this regard. Intergroup communication and expressed interest by top management are also helpful to a working development group in maintaining perspectives and morale.

Many of these organizational objectives can be facilitated through structure. A more "organic" organizational structure has been proposed as best for small-scale, technically complex R&D. This structure has few hierarchical levels and effects control from interpersonal feedback rather than by prior plan. Also, decision making is decentralized within this more flexible structure. The alternative to this has been dubbed a "mechanistic" structure, which is more hierarchical, operates more from set plans, and has more centralized decision making. This more formal arrangement is found in larger R&D efforts and it is thought to be effective for less-complex technological projects. The general trend in industrial R&D, however, is toward the reduction of the size of corporate laboratories, which at least should open more possibilities for organic organization.

Regardless of the organizational structure, however, the evaluation of the effectiveness and productivity of R&D projects is essential to their continuing success. Peer review by other scientists and engineers along with evaluation by "stakeholders" (e.g. potential users) are necessary for the group to evaluate the effectiveness of their outcomes. These assessments are best put to work in a particular firm, by using quantitative indicators that can be compared with others in the same industry.

With this brief review of factors that influence technological change in programs of R&D, we can see that progress or stagnation can occur at the working level for any one of a number of different reasons. With appropriate policies in place, effective assistance can be given to the innovation process for new energy technologies, in conjunction with funding support. We have been limited here to the functioning of the R&D establishment as it is constrained to the budget allocations handed it by the decision makers in government or industry. Policies and allocations, however, first require the setting of priorities. Many if not most of the priorities have been determined by the political and policy processes that are discussed in the the last section of Chapter 11.

References

Anon. (1991). Congress heaps funds for research in have-not states [Washington report]. *Physics Today,* Feb.,77-78.

Anon. (1994). Florida magnet lab opens [Research News]. *Science,* 30 Sept., 2007.

Baldwin, W.L. & Scott, J.T. (1987). *Market Structure and Technological Change.* Harwood.

Blissett, M., (1972). *Politics in Science.* Boston, MA: Little, Brown.

Brooks, H. (1968). *The Government of Science.* Cambridge, MA: MIT Press.

Crease, R.P. & Samios, N.P. (1989). Cold fusion confusion. *The New Yorker Magazine,* 24 Sept., 34-38.

de la Mothe & Ducharme, L. (1990). *Science, Technology and Free Trade.* London: Pinter.

Fermi, L. (1954). *Atomi in Famiglie.* Verona: A. Mondadori. [*Atoms in the Family,* English edn. 1965.]

Florida, R. & Kenney, M. (1990). *The Break-through Illusion.* New York: Basic Books.

Freiberger, P. (1984). *Fire in the Valley: The Making of the Personal Computer.* New York: McGraw-Hill.

Gipe, P. (1995). *Wind Comes of Age.* New York: Wiley.

Gleich, J. (1993). *Genius, The Life and Science of Richard Feynman.* New York: Vintage Books.

GRC (Gordon Research Conferences) (1973). The Gordon Research Conferences, based at the University of Rhode Island at Kingston, are a series of over a hundred conferences run every summer, each focused on a research specialty within physics, chemistry, biology or medicine. At the end of each 4–5 day conference, a questionnaire is circulated to the participants to determine their opinions on how that conference was run. One of the questions asks if they agree with the statement "Conference business was conducted in an open and democratic fashion, with avoidance of predominating groups or cliques". The author, while engaged in research in plasma physics, attended a Gordon Research Conference in August of 1973 at the Tilton School, in Tilton, NH, at which a questionnaire question of this type was asked.

Hoffman, B. (1972). *Albert Einstein – Creator and Rebel.* New York: Viking Press.

Kidder, T. (1981). *The Soul of a New Machine.* Boston, MA: Little Brown.

Latour, R. & Wodgar, S. (1979). *Laboratory Life: the Social Construction of Scientific Facts.* Beverly Hills, CA: Sage.

Maglich, B. (1991). Can the nation afford not to pursue research on aneutronic nuclear power? *Physics and Society,* 20,4–6. Holt, R. (1991).Tokamak magnetic fusion and the first generation of fusion power. *Physics and Society,* 20, 6–7. A pair of articles presenting the arguments for and against maverick proposals for research on reactors utilizing nuclear-fusion reactions not involving neutrons. Letters following on the same subject in *Physics and Society,* 20, Oct.

Mansfield, E. (1968). *The Economics of Technological Change.* New York: Norton.

Marshall, E. (1994). Peer review – congress finds little bias in system. *Science,* 265, 863.

Mulkay, M. (1991). *Sociology of Science – a Sociological Pilgrimage.* Bloomington,

IN: Indiana University Press.

Mulkay, M.J. (1976). The mediating role of the scientific elite. *Social Studies of Science*, 6, 445–470.

Polanyi, M. (1951). *The Logic of Liberty*. Chicago, IL: University of Chicago Press.

Reid, T.R. (1984). *The Chip: How Two Americans Invented the Microchip and Launched a Revolution*. New York: Simon & Schuster.

Richter, M.W. (1972). *Science as a Cultural Process*. Cambridge, MA: Schenkman.

Roe, A. (1953). *The Making of a Scientist*. New York: Dodd Mead.

Rogers, E. & Larsen, J. (1984). *Silicon Valley Fever: Growth of High-Technology Culture*. New York: Basic Books.

Sarewitz, D. (1996). *Frontiers of Illusion – Science, Technology, and the Politics of Progress*. Philadelphia, PA: Temple University Press.

Snow, C.P. (1951). *The Masters*. New York: Scribner.

Snow, C.P. (1964). *Corridors of Power*. New York: Scribner.

Storer, N.W. (1966). *The Social System of Science*. New York: Holt, Rinehart & Winston.

Tornatzky, L.G. & Fleischer, M. (1990). *The Processes of Technological Innovation*. Lexington, MA: Lexington Books.

Watson, J.D. (1981). *The Double Helix*. London: Weidenfeld & Nicholson.

Zuckerman, H. (1977). *Scientific Elites: Nobel Laureates in the United States*. New York: Free Press.

Index

accessibility of alternative technologies 226
advanced batteries 158–61, 165
aerodynamic lift, wind turbine blades
    115–16
agricultural wastes as energy resource 80
alcohol fuels
    from hydrolysis of cellulose 81, 84–5
    market prospects of 86
    production costs of 85–6
    properties of 85
alternative energy technologies 7
amorphous semiconductor solar cells 52
anaerobic digestion 89–91

batteries
    advanced 158–61, 165
    cycle lifetime 158–9, 163–4
    deep discharge in 158, 160
    energy storage 156–60
    high temperature 158–9
    lead–acid 157–8, 163–4
    sodium-sulfur 158–9
    weight-specific characteristics 158–60,
        164
BIG/GT see biomass-integrated gasifier/gas
    turbine
biochemical production of fuel, see alcohol,
    methane
biofuels
    gaseous 87–8
    impact of mass production 102–5
    liquid 82–3
    low-BTU gas 93, 95
    medium-BTU gas 88, 91, 93
    natural gas 4
    occupational health issues 104–5
    pipeline quality 87, 95
    producer gas 95
    prospects for 101–5
    pyrolysis 92–3
    town gas 93
    unit costs 98, 100–1
    US DOE program 87

see also alcohol, biomass, energy crops,
    methane
biogas gasifiers 88–92
    biomass-integrated gasifier/gas turbine
        95–7
    fixed bed 94
    fluidized bed 94
    integrated co-generation 95
    gasification 93–8
    Lurgi process 93–4
    municipal solid wastes 81–2
    pyrolytic 92–4
biomass
    carbon cycle 69–70
    charcoal 92
    from coal 102
    digester 88–92
    DOE Power Program 97
    as renewable resource 67
    short-rotation intensive culture trees
        70–1
    short-rotation woody crops 99
    wood 70–2
    wood fueled co-generation 68–9
    see also biogas digesters, energy crops
breakthrough development 272–4
building-integrated PV installations 58

capacity credit for wind power 128, 131
carbon cycle 69–70
carbon dioxide emissions 3
cellulosic energy crops 78–80, 84–5, 227
    hydrolysis of 81, 84–5
charcoal, historic use of 92
cold fusion 204
corn and sugar energy crops 83–84
corporate structure of research 234,
    269–272
costs
    analysis for energy 251–2
    competitive unit costs 225
    energy crops 85–6
    energy storage 12, 160–7, 175, 227

costs (*cont.*)
  geothermal   190
  hydropower   147–8
  photovoltaics   55–7
  R&D funding   245–51
  solar domestic hot water   24
  solar hydrogen   211–12
  tidal   195–6
  unit costs energy   11, 261
  wave energy   194
  wind   118, 120, 130–2
  wood   73–4
  world fuels   225
cross elasticities between fuels   225
crystalline semiconductor solar cells   50
cycle lifetime of batteries, *see* batteries,
      cycle lifetime

dedicated feedstock supply systems
      (DFSS)   96
development phase   236–8
direct solar electricity   47
DOE
  Biofuels Program   87
  Biomass Power Program   97

electric vehicle
  feasible applications   166
  driving range   158, 164, 227
  performance   164–6
  see also energy storage
electrolysis
  hydrogen conversion   169, 209–11
  hydrogen production   209–10
energy consumption demand   3
energy crisis, 1970s   4
energy crops   74–7
  agricultural waste   80
  biomass as renewable source   67
  cellulosic   78–80, 84–5, 227
  corn and sugar   83–4
  dedicated feedstock   96
  environmental impact   102–5, 227
  grasses   78–9
  herbaceous   78–9, 99
  land resources worldwide   76–8
  mass production, impact of   102–5
  processing   81, 86
  production
    costs   85–6
    potential   76–8
  species improvement   104

starchy   79–81
  see also biofuels, biomass
energy economy, transition   12
energy efficiency scenarios   3
energy R&D establishment   238–45
  mission   248–9
energy storage
  as aid to market penetration   163
  batteries   156–60
  economics   12, 163, 175
  electric energy   160–7
  fixed charges   162–4
  flywheels   167–8
  high temperature   176–7
  hot air   173
  hot water   172–3, 175
  hydrogen   168–72
  latent heat compounds   173–4
  mobile applications   156, 169–72
  operational availability from   160–2
  period displacement   162
  phase-change materials   173
  primary   153, 169
  role with sustainable sources   153–4,
      226
  seasonal   175–6
  secondary   153–68
  solar for industry   177
  stationary applications   153–6
  superconducting magnetism   179–81
  thermal   172–8
  ultra-capacitors   159–60
  see also batteries, electric vehicles,
      fuel-cell vehicles; *individual
      technologies*
environmental impact
  benefit of energy crops   103
  carbon dioxide emission   3
  fossil fuel use   3
  global warming   4
  greenhouse gases   3
  massive biofuels production   102–5, 227
  unanticipated   227
EV, *see* electric vehicle

FCV, *see* fuel-cell vehicle
flash pyrolysis   93
fluid-bed gasifiers   94
flywheels
  electromechanical storage   167–8
  high-speed low-mass   167–8
focused photovoltaic collectors   53

fossil fuels
  emissions 3
  exhausibility 4
fuel-cell vehicle 169–72, 217, 227

gaseous biofuels 87–8
  low BTU 93, 95
  medium BTU 88, 91, 93
  pipeline quality 87, 95
  see also methane, biofuels
gasification production of biofuels 93–8
geopressurized dome formations, heat
    resources in 188
geothermal energy 188–91
  heat and electricity costs 190
  heat applications 190
  power plants 188–90
  resources, worldwide 189–90
global warming 4
grasses as energy crops 78–9
greenhouse gases 3
grid-connected photovoltaic sites 59–60
  market penetration 60, 226

heat storage, *see* energy storage, solar
    domestic hot water
HECP, *see* herbaceous energy crop
    production
herbaceous energy crops 78–9
  production costs 99
high-temperature applications
  batteries 158–9
  heat storage 176–7
  superconductors 179–80
high-voltage transmission for large-scale
    power projects 146–8
hot air
  energy storage 173
  solar heating 27
hot water storage 172–3, 175
hot-gas cleanup for biomass gasification
    94, 96
hydroelectricity
  costs 146–7
  current status of 136–8
  generating capacity worldwide 137
  impact of 138–9, 141, 149, 228
  large-scale projects 144, 146–7
  remote site transmission 144–8
  turbines 142–4
  see also hydropower
hydrogasification 93

hydrogen
  from biomass 98, 215
  economy, the 207, 209, 217, 228
  end uses 216–17
  energy storage for mobile uses 168–72
  fuel cells 216–17
  fuel, history of 207–9
  from methane 98
  mobile applications 169–72
  photochemical production of 215–16
  pipelines 214–15
  production 209–11
  R&D program 218
  solar 211–13, 216
  storage costs 169
  from wind generation 213–14
hydrolysis cellulosic matter 81,
    84–5
hydropower
  capital costs of 147
  civil construction 144
  complementarity of sources 144
  development, potential for 149–50
  future of 148
  small-scale projects 139–42
  transmission costs 144–8
  see also hydroelectricity

induction generation, *see* wind turbines,
    induction generation
industrial process heat
  solar 19, 38
  storage for 177
inexhaustiblity 8–9
innovations, theories of 7
integrated co-generation using biomass 95
IPH, *see* industrial process heat

land resources for energy crops 74–8
landfills, methane derived from 91–2
latent-heat compounds, energy storage in
    173–4
lead–acid battery 157–8
liquid biofuels 82–3
load–demand matching 40, 59, 127–8
logistics curves 5
Lurgi process 93–4

mass production prospects, photovoltaics
    56–7
methane fuels 86, 88–92
  anaerobic production 89–91

methane fuels (*cont.*)
    biochemical production   88–92
    from landfill   91–2
    production costs   92
    steam reforming   98
    synthesis gas as source   97
methanol production costs   98
mission for energy R&D   248–9
MSW, *see* municipal solid wastes
multijunction solar cells   53
municipal solid wastes as energy resources
        81–82

National Phototvoltaic Program (USA)
        55
natural gas resources   4
nuclear fusion
    backstop resource   200
    breakeven point   201
    controlled   200
    criteria for sustainability   200
    inertial confinement of   202–3
    magnetic confinement of   201–2
    R&D, history of   201
nuclear waste   9

occupational health concerns   104–5
ocean energy, *see* ocean thermal, tidal
        power, wave power
ocean thermal energy conversion   196–8
    demonstration sites   197
    engineering challenges   198
    thermodynamics of power plants,   197
oil
    consumption, world   5
    market, world   4
    supplies security   4
operating availability of energy, *see* period
        displacement
OTEC, *see* ocean thermal energy
        conversion

p-n junctions   47–48
passive solar heating and cooling   29–32,
        177–8
PCM, see phase-change materials
period displacement   160–2
Petroleum Exporting Countries (OPEC)   4
phase-change materials, heat storage in
        173
photovoltaic systems
    building integrated   58

cells   48
focused collectors   53
grid connection   59–60
grid market penetration   60, 226
levelized costs of   55–7
market   57–60
mass production   56–7
National Photovoltaic program (USA)
        55
remote site   49–51, 58
textured cells   55
pipeline quality gaseous fuels   87, 95
politics of R&D funding   245–51
polycrystalline semiconductor solar cells
        50–2
primary energy storage   153, 169
producer gas   93
production costs, *see* costs
        see also individual fuels
public opinion
    policy for biomass fuel alternatives
        102
    support for energy R&D   229, 238–43
    wind farms   125–7
PV, *see* photovoltaic
pyrolysis production of biofuels   92–3

research and development, *see* R&D
R&D
    breakthrough development   272–4
    corporate structure   234, 269–72
    management   245, 269–76
    politics of funding   245–51
    processes   232–8, 269–76
    research   234–6
    stages   232–4
    strategic research   236
    strategies for energy crops   110
    support   229, 232, 239–43
recyclability   8–9
regional confinement of renewable
        resources   228
remote site installations   49–51, 58
renewability   8
    energy sources   8, 10
research *see* R&D
rural land use for wind farms   126

SDHW, *see* solar, domestic hot water
seasonal heat storage   175–6
secondary energy storage   153, 168
SEGS projects   39–42

semiconductor cells   50–2

short-rotation, intensive-culture trees
      70–71

short-rotation woody crop production
      99

SMES, *see* superconducting magnetic
      energy storage

sodium-sulfur batteries   158–9

solar
   cells
      amorphous semiconductor   52
      crystalline semiconductor   50
      multijunction   53
      polycrystalline semiconductor   50–2
      semiconductors   52
      thin film   52–3, 55–7, 226
   central receivers   35–7
   collectors
      bowls   45–6
      of hot water   20–1
      panels   49
      parabolic dish   34–5
      parabolic trough   33, 36, 38–42
      ponds   42–5, 177
   concentrators   32–42
   cooling, active   28–9
   cooling, passive   29–32, 177–8
   domestic hot water   19–26
      active systems   20
      circulation systems   20
      collection system   20, 22
      economics   24
      fraction   22
      market penetration   23–6, 226
      storage   22, 172
   electricity
      direct   47–8
      see also photovoltaics
   hydrogen   211–13, 216
      costs   211–12
   Solar One, Solar Two projects   35–8
   power
      for industrial process heat   177
      load–demand matching   40, 59, 127–9
      stand-alone costs   160
   space heating   26–8

SRIC, *see* short-rotation intensive-culture
      trees

SRWCP, *see* short-rotation woody crop
      production

starches/sugars for ethanol production
      83–4

starchy energy crops   79–81

steam reforming of methane   98

steam–carbon reaction   93

storage, *see* energy storage

strategic research   236

superconducting magnetic energy storage
      179–81

sustainability, defined   8

sustainable development   6, 8

sustainable energy sources   6, 8–9

sylvaculture practices   79–80

synthesis gas for methane production   97

tax incentives   25–6

technological innovation   238, 274–6

technological literacy   10

textured PV cells   55

thermal energy storage   172–8

thermochemical production of biofuels
      92

thin-film solar cells   52–3, 55–7, 226

Third World energy demands   5

tidal power   195–6
   capacity factors for plants   195
   costs   195–6
   environmental impact   196
   potential sites   196
   worldwide locations   195

town gas   93

transition, energy economy   12

tree plantations for fuel   70–4

ultra-capacitors   159–60

unanticipated outcomes   227

unit costs, energy   11, 261

VSCF, *see* wind turbines, variable
      speed/constant frequency

waste-to-energy projects   82, 91

wave energy
   costs of   194
   regional availability of   193–4
   remote-site applications of   193–4

weight-specific characteristics for batteries
      158–160, 164

wind (statistical) duration   124

wind power
   availability   121–5
   capacity costs   118, 120, 130–1
   capacity credit   128, 131
   economics of installations   131

wind power (*cont.*)
 energy output costs   120, 132
 farms   120
 history of   113–15
 induction generation   128–9
 load-demand matching   127–8
 market penetration   130–2, 226
 medium-term prospects for   129–30
 public incentives for   119–20
 R&D program   132
 regional potentials   123–6
 resources   121–9
  classes   123–5
 rural land use for   126

 turbines   114–17
  aerodynamic lift   115–16
  operating experience   117–21
  operating principles   114–17
  size considerations   118, 120, 127, 129
  variable speed/constant frequency   129
  vertical-axis   117, 119
wood
 co-generation with   68–9
 feedstock for biomass fuel   70–2
 sylvaculture   79–80
 waste as fuel   80

Printed in the United Kingdom
by Lightning Source UK Ltd.
107819UKS00002B/131-174

9 780521 018371